T5-CAO-232

Insecticides of Plant Origin

ACS SYMPOSIUM SERIES 387

Insecticides of Plant Origin

J. T. Arnason, EDITOR
University of Ottawa

B. J. R. Philogène, EDITOR
University of Ottawa

Peter Morand, EDITOR
University of Ottawa

Developed from a symposium sponsored
by the Division of Agrochemicals
of the American Chemical Society
at the Third Chemical Congress of North America
(195th National Meeting of the American Chemical Society),
Toronto, Ontario, Canada,
June 5-11, 1988

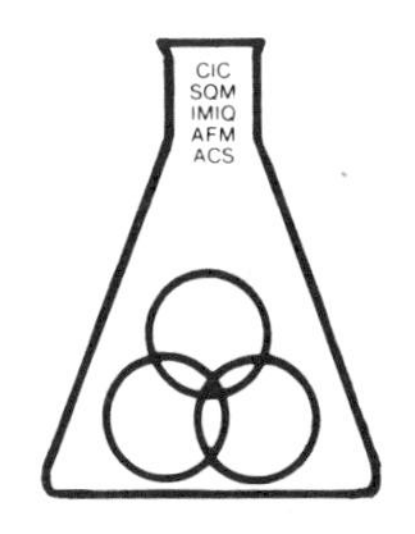

American Chemical Society, Washington, DC 1989

Library of Congress Cataloging-in-Publication Data

Insecticides of plant origin.
(ACS Symposium Series; 387)

"Developed from a symposium sponsored by the Division of Agrochemicals of the American Chemical Society at the Third Chemical Congress of North America (195th National Meeting of the American Chemical Society, Toronto, Ontario, Canada, June 5–11, 1988)."

Includes bibliographies and indexes.

1. Botanical insecticides—Congress.

I. Arnason, J. T. II. Philogène, B. J. R. III. Morand, P. (Peter) IV. American Chemical Society. Division of Agrochemicals. V. Series.

SB951.54.I57 1989 632'.951 89–85
ISBN 0–8412–1569–3
CIP

PRINTED IN THE UNITED STATES OF AMERICA

ACS Symposium Series

M. Joan Comstock, *Series Editor*

Foreword

The ACS SYMPOSIUM SERIES was founded in 1974 to provide a medium for publishing symposia quickly in book form. The format of the Series parallels that of the continuing ADVANCES IN CHEMISTRY SERIES except that, in order to save time, the papers are not typeset but are reproduced as they are submitted by the authors in camera-ready form. Papers are reviewed under the supervision of the Editors with the assistance of the Series Advisory Board and are selected to maintain the integrity of the symposia; however, verbatim reproductions of previously published papers are not accepted. Both reviews and reports of research are acceptable, because symposia may embrace both types of presentation.

Contents

Preface

PLANT-DERIVED SUBSTANCES have been used as botanical pesticides since ancient times. They were largely abandoned during the era of synthetic pesticides, but now the study of natural pesticides contributes novel approaches in control strategies for pests.

Botanical pesticides are plant natural products that belong to the group of so-called secondary metabolites, which includes thousands of alkaloids, terpenoids, phenolics, and minor secondary chemicals. These substances have no known function in photosynthesis, growth, or other basic aspects of plant physiology; however, their biological activity with insects, nematodes, and phytopathogenic fungi, among other organisms, is well-documented in the recent explosion of literature in chemical ecology. Selective pressure exerted by plant pests has probably contributed to the evolution of natural protective agents designed to disrupt pest physiology. Apparently, almost every plant species has developed a unique chemical complex that protects it from pests. Thus, the plant kingdom offers us a diverse group of complex chemical structures and almost every imaginable biological activity.

The discovery rate for new insecticides from synthetic sources has declined in recent years. Furthermore, synthetic insecticides that share a neurotoxic mode of action can lead to the development of cross resistance in plants. Fortunately, we can find many new alternatives within the large group of natural substances. In addition, many botanical pesticides have the advantage of providing novel modes of action that reduce the risk of cross resistance. Naturally occurring mixtures of substances provide a multifactoral selective pressure that slows the development of resistance in pests. More important, researchers in this area have discovered subtle but effective mechanisms of pest control, such as the behavior-modifying antifeedants, repellents, and oviposition deterrents.

The symposium on which this book is based brought together specialists from Africa and Asia, Europe and North America. Among the chapters in this volume are historical, ethnobotanical reports about the use of plants by native North Americans and accounts of traditional use of plants by African and Asian farmers. Martin Jacobson, who has the longest association of the authors with the field, describes the modern history of botanicals and surveys the range of species investigated in recent years. Among the new botanical pesticides examined in this volume are acetogenins of the *Annonaceae,* agarofurans of the *Celastraceae,* and limonoids of the *Meliaceae.* The authors deal with the metabolism and toxicity of acetylchromenes in insects, electrophysiological bases of antifeedant action of nonprotein amino acids, and the repellent action of eugenol to stored grain pests.

Several chapters focus on the botanical pesticide, neem, which is close to registration and has potent antifeedant and molt-inhibiting properties and can move systemically in plants. Neem is used worldwide and has a preliminary registration in the United States. Authors evaluate its mode of action and performance under field conditions. The novel phototoxic thiophenes, which are exceptional larvicides of the malaria mosquito, are described. Pipercicides are the subject of a chapter about synthetic approaches to developing highly active insecticides from a promising lead. Transgenic technologies that use petunia sterones are explored.

Acknowledgments

The symposium and book would have been impossible without the support of a major grant from the Natural Sciences and Engineering Research Council of Canada and generous support from Cyanamid Canada Ltd., Safer Canada Ltd., the Chemical Institute of Canada, the Canadian Society for Chemistry, and the American Chemical Society. We are especially grateful to the many reviewers who contributed their time and effort and to J. Pierce, who suggested the project.

J. T. ARNASON
B. J. R. PHILOGENE
PETER MORAND
University of Ottawa
Ottawa, Ontario K1N 6N5, Canada

December 14, 1988

Chapter 1

Botanical Pesticides

Past, Present, and Future

Martin Jacobson[1]

1131 University Boulevard West, Apartment 616, Silver Spring, MD 20902

Until the Second World War, the only botanical pesticides used in the Western Hemisphere were pyrethrum, rotenone, nicotine, sabadilla, and quassin. Rotenone is presently used only on a limited number of crops due to its high toxicity to fish, and natural pyrethrum from chrysanthemum flowers is employed mainly as a rapid knockdown agent for crawling and flying insects affecting man and animals. Nicotine, sabadilla, and quassin are seldom used as pesticides today. Based on my 44 years of personal research experience with several thousand plants of many families as pesticides,it is my opinion that the most promising botanicals for use at the present time and in the future are species of the families Meliaceae, Rutaceae, Asteraceae, Malvaceae, Labiatae, and Canellaceae; their research and applicability are reviewed.

From early Roman times to the mid-20th century, only pyrethrum, rotenone, nicotine, sabadilla and quassin were widely used as insect repellents and toxicants in the Western Hemisphere. The discovery of DDT, initially believed to be the universal synthetic panacea from insect attack, proved after prolonged use to be disastrous. Rachel Carson's "Silent Spring" in 1962 signalled DDT's practical elimination due to its extreme persistence, toxicity, bioaccumulation, and tendency to cause malignancy. This led to a frantic search for "safer" synthetics, culminating in the chlorinated hydrocarbons (aldrin, dieldrin, chlordane, and heptachlor). These compounds also proved to be unduly toxic, ecologically disastrous, or induced insect resistance, and their use has either been severely restricted or banned by the Environmental Protection Agency (EPA). In the search for safer alternatives, attention has once again turned to botanicals.

Rotenone is presently used only on a limited number of crops due to its high toxicity to fish, natural pyrethrum from chrysanthemum flowers is employed mainly as a rapid knockdown agent for crawling and flying insects affecting man and animals, and nicotine, sabadilla and quassin are rarely used today.

The most promising botanicals for use at the present time and in the future are species of the families Meliaceae, Rutaceae, Asteraceae, Annonaceae, Labiatae, and Canellaceae.

[1]Retired from U.S. Department of Agriculture.

0097–6156/89/0387–0001$06.00/0

MELIACEAE

The neem tree, Azadirachta indica A. Juss (synonym Melia azadirachta A. Juss), also known commonly as "nim" or "margosa." is a sub-tropical tree native to the arid areas of Asia and Africa. It is presently being cultivated in a number of Central American and South American countries as well. Extracts of various parts of the tree, but especially of the seeds, have been shown to possess feeding deterrency, repellency, toxicity, and growth disruptive properties to numerous species and stages of insects of many orders, and this has earned for the tree its reputation as the "wonder tree". A comprehensive review of the entomological properties of neem has been published (1).

A host of tetranortriterpenoids have been isolated and identified from various parts of the tree, but the major entomologically active component is azadirachtin, whose highly complicated structure was first reported by Zanno (2) and Nakanishi (3) and most recently revised from the stereochemical standpoint (4,5). Isomeric azadirachtins (A-G) have been isolated by Forster (6) and Rembold et al (7,8).

Azadirachtin is effective as a feeding deterrent, repellent, toxicant, sterilant, and growth disruptant for insects at dosages as low as 0.1 ppm. Homemade formulations of neem seeds have been used by farmers in the developing countries for many years and are receiving even greater use today. Although the molecular complexity of azadirachtin probably precludes its synthesis in the near future, both crude and partially purified extracts can be used for pest control and have been shown to be safe for man and animals (Jacobson, M.; Ed., The Neem Tree, CRC Press, Boca Raton, Florida, in press).

A commercial patent (9) has been issued describing the neem seed formulation known as "Margosan-O" for use on nonfood crops and in nurseries, as approved by EPA, and the improved preparation and marketing of neem formulations in the United States seems to be assured.

The Chinaberry tree, Melia azedarach L., is frequently mistaken in the literature for the neem tree. Although it occurs in much the same areas of the world, its seeds contain several of the limonoids (10,11) common to neem seeds except azadirachtin (10) and its extracts are insecticidally active (Larson, R.O., personal communication), chinaberry does not have a bright future as a pesticide due to its extreme toxicity to warm-blooded animals.

Cedrela Odorata L., Spanish cedar. Ethanol extracts of the leaves roots, root bark, or twigs of this tree are reported to prevent feeding by adult striped cucumber beetles, Acalymma vittatum (Fabricius) in the greenhouse. Also effective in this regard are extracts of the twigs, leaves, or roots of C. Toona Roxb. ex Rottl. & Willd, the root bark of Carapa guianensis Aubl., and the leaves of Trichilia hispida (Reed, D.K., personal communication).

Trichilia roka L. Limonoids (trichilins A-F) in the root bark are known to be antifeedants for larvae of Spodoptera eridania (Cramer)(southern army worm), and adult Epilachna varivestis Mulsant (Mexican bean beetle) (12). The limonoid sendanin, isolated from the fruits (13), is a potent growth inhibitor for Heliothis virescens Fabricius (tobacco budworm), Spodoptera frugiperda (J.E. Smith)(fall armyworm). and H. zea (Boddie)(corn earworm)(14).

Trichilia hispida Penning. Limonoids (hispidins A-C) were isolated from this species (15-16).

Carapa procera DC. Three tetranortriterpenoids, designated "carapolides A, B, and

C," were isolated from the seeds (17), and "carapolides D, E, and F" were isolated from the seeds of C. grandiflora Sprague (18).

Guarea cedrata L. Three tetranortriterpenoids have been isolated from the bark of this tree (19).

Swietenia macrophylla King
S. mahagoni DC,mahagony. The tetranortriterpenoids swietenine and swietenolide have been isolated from the seeds of these species (20,21). 2-Hydroxyswietenine has also been obtained from mahagony seeds (22,23).

Toona ciliata Roemer, synonym Cedrela toona Roxb. ex Rottl. & Willd. Although Spanish cedar, Cedrela odorata, is heavily attacked in Central America by larvae of the moth Hipsipyla grandella Zeller, the Australian variety, C. toona var. australis, is highly resistant to attack by this pest (24-27).

Kraus et al. (28,29) isolated from the leaves of the Australian variety two β-secotetranortriterpenoids, which they named "toonacilin" and "6-acetoxytoonacilin", respectively, that showed strong antifeeding and insecticidal activity against H. grandella and the Mexican bean beetle. These compounds, as well as two additional tetranortriterpenoids, were subsequently isolated from the bark of the tree and their structures were determined (28).

Toona sureni (Blume) Merrill. Kypke (30) isolated, from the bark and leaves of this tree, toonacillin and a new triterpenoid designated "surenolactone". This proved to be the first tetranortriterpenoid-A/B-dilactone found in a meliaceous plant.

Melia dubia Cav., synonym M. compositae Willd. Two new triterpenoids have recently been isolated from the leaves and seeds of this species. They have been named "compositin" (1,7-ditiglylvilasinin) and "compositolide" (31,32). However, the biological activity of these compounds has not yet been reported.

Melia volkensii Guerke. Mwangi (12) reported that an aqueous extract of the fruit kernels of this tree from East Africa showed antifeedant activity against nymphs and adults of Schistocerca granaria (Forsk.)(desert locust) when incorporated into the diet. The isolation and structure determination of a new limonoid, "volkensin" and the previously known salannin have very recently been reported and proved to be deterrent to larvae of the fall armyworm (Rajab, M.S., personal communication).

RUTACEAE

Arnason et al. (33) evaluated six limonoids from the families Rutaceae and Meliaceae as feeding deterrents for the European corn borer, Ostrinia Nubilalis Hübner, following incorporation into the diet at 50 ppm. The order of activity was gedunin > bussein > entandrophragmin > nomilin > cedrelone > anthothecol.

Limonin, a limonoid isolated from several citrus species of the family Rutaceae (especially orange and grapefruit) was ten-fold less active than nomilin, obacunone, and azadirachtin as a feeding deterrent for Heliothis zea (34). Nomilin was almost as active as azadirachtin. All of these limonoids were at least ten-fold more active against fall armyworm larvae than against H. Zea. Azadirachtin was the most active (35).

An excellent review covering the isolation of citrus limonoids by preparative high performance liquid chromatography (HPLC) is that by Rouseff (36).

Limonoids such as methyl deacetylnomilinate and methyl isolimonate were isolated from Citrus reticulata Blanco var. austera (37) and C. aurantium L. (38), respectively.

Limonoids from Citrus paradisi Macfadyen (19) prevents feeding by Spodoptera litura and nomilin prevents feeding by S. frugiperda and trichoplusia ni Hübner (cabbage looper) (39). Limonin is also an effective antifeedant for Leptinotarsa decemlineata (Say)(Colorado potato beetle)(40) but not against Spodoptera exempta (Walker)(beet armyworm), Maruca testularis (Geyer)(bean podborer), or Eldana saccharina L. (41).

Citrus oils have proved to be toxic and deterrent to several species of stored product insects such as Callosobruchus maculatus (Fabricius)(cowpea weevil) and Sitophilus oryzae (Linnaeus)(rice weevil)(42-44).

Tecleanin, a possible precursor of limonin, and several other tetranortriterpenoids have been isolated from the stem bark of Teclea grandifolia Engl. (45).

A number of terpene hydrocarbons, especially limonene, isolated from the leaves of Orixa japonica Thunb. and Clausena anisata (Willd.) Oliv. root bark deter feeding by Spodoptera litura and are toxic to Callosobruchus phasecoli Gyll. (bean weevil)(46-48).

Zanthophylline, an alkaloid isolated from the stems and branches of Zanthophyllum monophyllum L., deterred feeding by Hemileuca oliviae Cockerell (range caterpillar, Melanoplus sanguinipes (Fabricius)(migratory grasshopper), Hypera postica (Gyllenhal)(alfalfa weevil) and Schizaphis graminum (Rondani)(greenbug)(49). Three 9-acridone alkaloids isolated from Teclea trichocarpa Eng. bark have been shown to deter feeding by the beet armyworm (50). The alkaloid N-methylflindersine and several benz(C)-phenanthridine alkaloids obtained from the East African root bark of Fagara chalybea Engl. and F. holstii F. deterred feeding by beet armyworm and the Mexican bean beetle (51). Five indoloquinazoline alkaloids isolated from the ripe fruits of Evodia rutaecarpa Hook f. & Thoms. inhibited the growth of Bombyx mori (Linnaeus)(silkworm) larvae (52).

Herculin, a pungent isobutylamide (N-isobutyl-2,8-dodecadienamide), was isolated from the bark of southern prickly ash, Zanthoxylum clava-herculis L., and shown to be as toxic as the pyrethrins to Musca domestica (Linnaeus)(house fly), mosquito larvae, and ticks, and ovicidal to Pediculus humanus humanus Linnaeus (body louse)(53). However, the compound is quite unstable owing to its unsaturation. A number of related unsaturated isobutylamides isolated from plants of the family Asteraceae show the same type of activity and rapid knockdown of flying insects (54).

ASTERACEAE

Sesquiterpene lactones isolated from a number of species of this family have proved to be excellent feeding deterrents for pest insects.

The germacranolides, schkuhrin-I and schkuhrin-II, isolated from the whole plant of Schkuhria pinnata (Lam.) O. Kuntze, exhibit antifeedant activity against the beet armyworm and the Mexican bean beetle, antibacterial activity against some Gram-positive organisms, and cytotoxicity (carcinostats) (55).

The sesquiterpene lactone, alantolactone, isolated from Inula helenium L., significantly reduces feeding and survival of Tribolium confusum Jacquelin du Val. (confused flour beetle) (56).

Isoalantolactone deters feeding by Sitophilus granarium (Linnaeus)(granary weevil), confused flour beetle, and Trogoderma granarium Everts (khapra beetle)(57). Sesquiterpene lactones such as bisabolangelone and related compounds also affect these species (58) and the beet armyworm (59,60).

In 1978, Maradufu et al. (61) isolated the simple ketone, 5-ocimenone, from the leaves of the Mexican marigold, Tagetes minuta L., which is responsible for the repellency of the plant oil to adult Anopheles mosquitoes and for its toxicity to the mosquito larvae (62). The oil also possesses juvenilizing properties against Dysdercus koenigii F. (63).

Achillea millefolium L. is the source of N-(2-methylpropyl)-E2, E4-decadienamide, a strong larvicide for Aedes mosquitoes (64), and Arnason et al. (65) have reported on a number of mosquito larvicidal acetylenes obtained from plants of the Asteraceae.

In a survey of the effect of ethereal extracts of plants from many families, Jacobson et al. (66, 67) were able to demonstrate that the roots of the American coneflower, Echinacea angustifolia DC, contain components toxic to Aedes mosquito larvae and house flies and disruptive to the growth and development of the yellow mealworm, Tenebrio molitor Linnaeus, and the large milkweed bug, Onchopeltus fasciatus (Dallas). Echinacein, the toxic component, was identified as N-isobutyl-(E,Z,E,E)-2,6,8,10-dodecatetraenamide (68). Juvenilization is due to "echinolone," identified as 4,10-dimethyl-10-hydroxy-(E)-4,11-dodecadien-2-one (69).

Extremely high juvenilizing activity on several species of insects has been shown by 6,7-dimethoxy-2,2-dimethylchromene ("precocene II"), isolated from the bedding plant, Ageratum houstonianum Mill. (70-73).

ANNONACEAE

Ethereal extracts of the twigs of Annona senegalensis Pers. are highly toxic to the large milkweed bug (67), and ethanolic extracts of the fruit of the custard apple, A. reticulata L., A. glabra L., and A. purpurea L. fruit exert a severe juvenilizing effect on the striped cucumber beetle adults, Diabrotica sp. (Reed, D.K., unpublished results). One or several of the kaurane and 16-kaurene diterpenes isolated from the stem bark of these species (74) may be responsible for the activity.

Numerous alkaloids have been reported from several species of Annona. One of these, liriodenine, isolated from A. glabra stem bark, is an active tumor (KB) inhibitor (75), as are four other alkaloids obtained from the stems and leaves of A. purpurea (76). Annonelliptine, an isoquinoline alkaloid isolated from A. elliptica R.E. Fries, is also suspected to be involved in this activity (77).

MALVACEAE

It was long thought that gossypol, present in cotton varieties, Gossypium hirsutum L., resistant to the bollworm, Spodoptera littoralis (Boisd.), was the major or even sole component responsible for the resistance (Stipanovic, R.D., personal communication). It is now known that, in addition to gossypol, "other related terpenoids, condensed tannins....and certain monomeric flavonoids" also provide resistance to the tobacco budworm, the corn earworm, and the pink bollworm, Pectinophora gossypiella (Saunders) (78). The resistance of the Rose-of-Sharon (Hibiscus syriacus L.) to feeding by the boll weevil, Anthonomus grandis Boheman, is known to be directly connected with a mixture of unsaturated fatty acids and their methyl esters (79). However, the value of high gossypol content in cotton strains for suppressing the bollworm and the spiny bollworm, Earias insulana (Boisd.), was demonstrated in unsprayed fields in Israel (80). In addition, synergistic and antagonistic interactions of gossypol with some organic phosphorus insecticides were demonstrated in the bollworm by topical application (81).

Interestingly, gossypol is presently under study in several parts of the world as an antifertility agent for the human male and is now being used for this purpose in China. Large scale clinical trials show that this compound, taken orally in pill form by male animals and humans, is safe, effective, inexpensive, and reversible in its effects (inhibiting sperm production) (82).

LABIATAE

The leaf oil of sweet basil, Ocimum basilicum L., contains clerodanes known as "juvocimene-I" and "juvocimene-II" which have a juvenilizing effect on the milkweed bug, Oncopeltus fasciatus (Dallas), when applied topically (83,84). The oil is also a powerful larvicide for Culex mosquitoes; this action is probably also due to the juvocimenes, although the content of linalool and several sesquiterpenes may play a significant role (85).

Following the disclosure, in 1980 and 1981, of the feeding deterrent and juvenilizing effects of Ajuga reptans L. and A. remota Benth. on the Mexican bean beetle (86) and the two-spotted spider mite, Tetranychus urticae Koch (87), the presence of numerous clerodane diterpenes was reported in A. reptans (88), A. chamaepitys (L.) Schreb. (89), A. nipponensis Makino (90,91), A. iva (L.) Schreb. (92), and A. pseudovia Robell. & Cast. ex DC (93). These extracts also prevent feeding by the bollworm and by Pieris brassicae L. (94-96) and cause juvenilization of the beet armyworm (97). The components mainly responsible for the biological activity are isomeric ajugarins which have been synthesized (98-103).

CANELLACEAE

A series of sesquiterpenoid dialdehydes (warburganals, polygodial, muzigadial) isolated from East African trees of the genera Warburgia and Polygonum have been shown to be powerful feeding deterrents against larvae of the beet armyworm, bollworm (Heliothis armigera L.)., and H. virescens (104-108), as well as adult striped cucumber beetles (Reed, D.K., unpublished results). Warburganal also possesses antiyeast and antifungal activity (109). Procedures for synthesizing polygodial (110-112), warburganal (113,114), and muzigadial (115) are available.

CONCLUSION

It is clear that, in my opinion, many plants of the families Meliaceae, Rutaceae, Asteraceae, Malvaceae, Labiatae, and Canellaceae are capable of being used as pesticides if certain conditions can be met. With the exception of neem, which has been shown to be safe, easily grown or available in large quantities, and is already in use as a pesticide in the developing countries (besides being approved for use on nonfood crops and in nurseries in the United States)(116), other genera and species of these families must meet the following criteria before they can be used commercially: (1) safe for plant and animal life, (2) biodegradable (environmentally safe), (3) ready availability of the plant or capability for cultivation, (4) determination of isolation procedures for the active component or components (or of formulation of extracts prepared from plant parts) or (5) establishment of synthetic procedures for the active components. (117).

Literature Cited

1. Jacobson, M. ACS Symp. Ser. 296, American Chemical Society, Washington, D.C., 1986, 220-32.
2. Zanno, P.R. Ph.D. Thesis, Columbia University, New York, 1974.
3. Nakanishi, K., Recent Adv. Phytochem. 1975, 9, 283-98.
4. Bilton, J.N.; Broughton, H.B.; Ley, S.V.; Lidert, Z.; Morgan, E.D.; Rzepa, H.S.; Sheppard, R.N. J. Chem. Soc., Chem. Commun. 1985, 968-71.
5. Kraus, W.; Bokel, M.; Bruhn, A.; Cramer, R.; Klaiber, I.; Klenk, A.; Nagl, G.; Poehnl, H.; Sadlo, H.; Vogler, B. Tetrahedron 1987, 43, 2817.
6. Forster, H. M.S. Thesis, University of Munich, FRG, 1983.
7. Rembold, H.; Forster, H.; Czoppelt, Ch.; Rao, P.J.; Sieber, K.P. In Natural Pesticides from the Neem Tree and Other Tropical Plants; Schmutterer, H. and Ascher, K.R.S., Eds., GTZ Press, Eschborn, FRG, 1984, p. 153.
8. Rembold, H. In Advances in Invertebrate Reproduction; Engels, W. et al., Eds., Elsevier; New York, 1984; Vol. 3, p 481.
9. Larson, R.O.; U.S. Patent 4 556 562, 1985.
10. Srivastava, S.D. J. Nat. Prod. 1986, 49, 56-61.
11. Lee, S.M.; Klocke, J.A.; Balandrin, M.F. Tetrahedron Lett. 1987, 28, 3543-6.
12. Mwangi, R.W. Entomol. Exp. Appl. 1982, 32, 277-80.
13. Nakatani, M.; Iwashita, T.; Naoki, H.; Hase, T. Phytochemistry 1985, 24, 195-6.
14. Nakatani, M.; James, J.C.; Nakanishi, K. J. Am. Chem. Soc. 1981, 103, 1228-30.
15. Kubo, I.; Klocke, J.A. Experientia 1982, 38, 639-40.
16. Jolad, S.D.; Hoffmann, J.J.; Cole, J.R.; Tempesta, M.S.; Bates, R.B. J. Org. Chem. 1980, 45, 3132.
17. Jolad, S.D.; Hoffmann, J.A.; Schram, K.H.; Cole, J.R.; Tempesta, M.S.; Bates, R.B. J. Org. Chem. 1981, 46, 641-4.
18. Kimbu, S.F.; Ayafor, J.F.; Sondengam, B.L.; Connolly, J.D.; Rycroft, D.S. Tetrahedron Lett. 1984, 25, 1613-16.
19. Kimbu, S.F.; Ayafor, J.F.; Sondengam, B.L.; Teamo, E.; Connolly, J.D.; Rycroft, D.S. Tetrahedron Lett. 1984, 25, 1617-20.
20. Akimniyi, J.A.; Connolly, J.D.; Rycroft, D.S.; Sondengam, B.L.; Ifeadike, N.P. East Afr. Med. J. 1980, 58, 1865-8.
21. Connolly, J.D.; Labbe, C. J. Chem. Soc., Perkin Trans. 1980, (2), 529.
22. Ghosh, S. J. Ind. Chem. Soc. 1960, 37.
23 Daily, A.; Seligmann, O.; Lotter, H.; Wagner, H. Z. Naturforsch. 1985, 40C, 519-22.
24. Grijpma, P.; Ramalho, R. Turrialba 1969, 19, 531-47.
25. Grijpma, P. Turrialba 1970, 20, 85-93.
26. Grijpma, P.; Roberts, S.C. Turrialba 1975, 25, 152-9.
27. Allan, G.C.; Gara, R.I.; Roberts, S.C. Turrialba 1975, 25, 255-9.
28. Kraus, W.; Grimminger, W.; Sawitski, G. Angew. Chem. (Intl. Ed. Engl.) 1978, 17, 492-3.
29. Kraus, W.; Grimminger, W. Nouveau J. Chem. 1980, 4, 651-5.
30. Kypke, K. Ph.D. Thesis, Hohenheim University, FRG, 1980.
31. Kraus, W.; Kypke, K.; Bokel, M.; Grimminger, W.; Sawitski, G.; Schwinger, C. Liebigs Ann. Chem. 1982, 87-98.
32. Koul, O.Z. Z. Angew. Entomol. 1983, 95, 166-71.
33. Arnason, J.T.; Philogene, B.J.R.; Donskov, N.; Kubo, I. Entomol. Exp. Appl. 1987, 43, 221-6.

34. Klocke, J.A.; Kubo, I. Entomol. Exp. Appl. 1982, 32, 299-301.
35. Anonymous. Citrus Vegetable Mag. 1982, (5), 32-3.
36. Rouseff, R.L. In Semiochemistry: Flavors and Pheromones: Acree, T.E.; Soderlund, D.M., Eds.: Walter de Gruyter, Berlin, FRG, 1985, p 275.
37. Bennett, R.D.; Hasegawa, S. Phytochemistry 1980, 19, 2417-19.
38. Bennett, R.D.; Hasegawa, S. Tetrahedron 1981, 37, 17-24.
39. Altieri, M.A.; Lippmann, M.; Schmidt, L.L.; Kubo, I. Prot. Ecol. 1984, 6, 91-4.
40. Alford, A.R.; Cullen, J.A.; Storch, R.H.; Bentley, M.D. J. Econ. Entomol. 1987, 80, 575-8.
41. Hassanalli, A.; Bentley, M.D.; Ole Sitayo, E.N.; Sjoroge, P.E.W.; Yatagai, M. Insect Sci. Appl. 1986, 7, 495-9.
42. Su, H.C.F.; Speirs, R.D.; Mahany, P.G. J. Econ. Entomol. 1972, 65, 1433-6.
43. Su, H.C.F. J. Georgia Entomol. Soc. 1976, 11, 297-301.
44. Su, H.C.F.; Speirs, R.D.; Mahany, P.G. J. Econ. Entomol. 1972, 65, 1438-41.
45. Ayafor, J.F.; Sondengam, B.L.; Connolly, J.D.; Rycroft, D.S.; Okogun, J.I. J. Chem. Soc., Perkin I Trans. 1981, 1750-3.
46. Yajima, T.; Kato, N.; Munakata, K. Agr. Biol. Chem. 1977, 41, 1263-8.
47. Mester, I.; Szendrei, K.; Reisch, J. Planta Med. 1977, 32, 81-5.
48. Taylor, W.E.; Vickery, B. Ghana J. Agr. Sci. 1974, 7, 61-2.
49. Capinera, J.L.; Stermitz, F.R. J. Chem. Ecol. 1979, 5, 767-71.
50. Lwande, W.; Gebreyesus, T.; Chapya, A.; Macfoy, C.; Hassanali; Okech, M. Insect Sci. Appl. 1983, 4, 393-5.
51. Chou, F.Y.; Hostettmann, K.; Kubo, I.; Nakanishi, K.; Taniguchi, M. Heterocycles 1977, 7, 169-77.
52. Kamikado, T.; Murakoshi, S.; Tamura, S. Agr. Biol. Chem. 1978, 42, 1515-19.
53. Jacobson, M. J. Am. Chem. Soc. 1948, 70, 4234-7.
54. Jacobson, M. In Naturally Occurring Insecticides: Jacobson, M. Crosby, D.G., Eds.: Dekker, New York, 1971; p 137.
55. Pettei, M.J.; Miura, I.; Kubo, I.; Nakanishi, K. Heterocycles 1978, 471-80.
56. Pirman, A.K.; Elliott, R.H.; Towers, G.H.N. Biochem. System Ecol. 1978, 6, 333-5.
57. Steibl. M.; Nawrot, J.; Herout, V. Biochem. System Ecol. 1983, 11, 381-2.
58. Harmatha, J.; Nawrot, J. Biochem. System Ecol. 1984, 12, 95-8.
59. Nawrot, J.; Harmatha, J.; Novotny, L. Biochem. System Ecol. 1984, 12, 99-101.
60. Ganjian, I., Kubo, I.; Fludzinski, P. Phytochemistry 1983, 2525-6.
61. Maradufu, A.; Lubega, R.; Dorn. F. Lloydia 1978, 41, 181-3.
62. Okoth, J. East Afr. Med. J. 1973, 50, 317-22.
63. Saxena, B.P.; Srivastava, J.B. Ind. J. Exp. Biol. 1973, 11, 56-8.
64. Lalonde, R.T.; Wong, C.F.; Hosftead, S.J.; Morris, C.D.; Gardner, L.C. J. Chem. Ecol. 1980, 6, 35-48.
65. Arnason, T.; Swain, T.; Wat, C.K.; Graham, E.A.; Partington, S.; Towers, C.H.N. Biochem. System Ecol. 1981, 9, 63-8.
66. Jacobson, M. Mitteil. Schweiz. Entomol. Ges. 1971, 44, 73-7.
67. Jacobson, M.; Redfern, R.E.; Mills, G.D., Jr. Lloydia 1975, 38, 455-72.
68. Jacobson, M. J. Org. Chem. 1967, 32, 1646-7.
69. Jacobson, M.; Redfern, R.E.; Mills, G.D., Jr. Lloydia 1975, 38, 473-6.

70. Bowers, W.S.; Ohta, T.; Cleare, J.S.; Marsella, P.A. Science Science 1976, 198, 542-7.
71. Brooks, G.T.; Pratt, G.E.; Jennings, R.C. Nature 1979, 281, 570-2.
72. Dees, W.H.; Sonenshine, D.E.; Freidling, E.; Buford, N.P.; Khalil, G.M. J. Med. Entomol. 1982, 19, 734-42.
73. Deb, D.C.; Chakravorty, S. J. Insect Physiol. 1982, 28, 703-12.
74. Eyse, J.T.; Gray, A.I.; Waterman, P.G. J. Nat. Prod. 1987, 50, 979-83.
75. Warthen, D.J., Jr.; Gooden, E.L.; Jacobson, M. J. Pharm. Sci. 1969, 58, 637-8.
76. Sonnet, P.E.; Jacobson, M. J. Pharm. Sci. 1971, 60, 1254-6.
77. Sandoval, D.; Preiss, A.; Schreiber, K.; Ripperger, H. Phytochemistry 1985, 24, 375-6.
78. Waiss, A.C., Jr. Agr. Res. (USDA) 1982, 30, (1), 4-5.
79. Bird, T.C.; Hedin, P.A.; Burks, M.L. J. Chem. Ecol. 1987, 13, 1087-97.
80. Zur, M.; Meisner, J.; Kabonci, E.; Ascher, K.R.S. Phytoparasitica 1980, 8, 189-94.
81. Meisner, J.; Ascher, K.R.S.; Zur, M.; Etzik, C. Z. Pflanzenkrankh. Pflanzenschutz 1982, 89, 571-4.
82. Murthy, R.S.R.; Basu, D.K.; Murti, V.V.S. Curr. Sci. (India) 1981, 50, 64-6.
83. Bowers, W.S.; Nishida, R. Science 1980, 209, 1030-2.
84. Nishida, R.; Bowers, W.S.; Evans, P.H. J. Chem. Ecol. 1984, 10, 1435-51.
85. Chavan, S.R.; Nikam, S.T. Ind. J. Med. Res. 1982, 75, 220-2.
86. Schmutterer, H.; Tervooren, G. Z. Angew. Entomol. 1980, 89, 470-8.
87. Schauer, M.; Schmutterer, H. Z. Angew. Entomol. 1981, 91, 425-33.
88. Camps, F.; Coll, J.; Cortel, A. Chem. Lett. (Tokyo) 1981, 1093-6.
89. Hernandez, A.; Pascual, C.; Sanz, J.; Rodriguez, B. Phytochemistry 1982, 21, 2909-11.
90. Shimamura, H.; Sashido, Y.; Ogawa, K.; Iitaka, Y. Tetrahedron Lett. 1981, 22, 1367-8.
91. Shimamura, H.; Sashido, Y.; Ogawa, K.; Iitaka, Y. Chem. Pharm. Bull. 1983, 31.
92. Camps, F.; Coll, J.; Cortel, A. Chem. Lett. (Tokyo) 1982, 1053-6.
93. Camps, F.; Coll, J.; Dargallo, O. Phytochemistry 1984, 23, 387-9.
94. Bellas, K.; Camps, F.; Coll, J.; Piulacks, D. J. Chem. Ecol. 1985, 11, 1439-45.
95. Abivardi, C.; Benz, G. Mitteil. Schweiz. Entomol. Ges. 1984, 57, 383-92.
96. Geuskens, R.B.M.; Luteijn, J.M.; Schoonhoven, L.M. Experientia 1983, 30, 403-4.
97. Kubo, I.; Klocke, J.A.; Asano, S. Agr. Biol. Chem. 1981, 45, 1925-7.
98. Kubo, I.; Kido, M.; Fukuyama, Y. J. Chem. Soc.; Chem. Commun. 1980, 897-8.
99. Kubo, I.; Klocke, J.A.; Miura, I.; Fukuyama, Y. J. Chem. Soc., Chem. Commun. 1982, 618-19.
100. Kende, A.S.; Roth, B. Tetrahedron Lett. 1982, 23, 1751-4.
101. Luteijn, J.M.; de Groot, A. Tetrahedron Lett. 1982, 23, 3421-4.
102. Ley, S.V.; Simpkins, N.S.; Whittle, A.J. J. Chem. Soc., Chem Commun. 1983, 503-5.
103. Jones, P.S.; Ley, S.V.; Simpkins, N.S.; Whittle, A.J. Tetrahedron 1986, 42, 6519-34.
104. Blaney, W.M.; Simmonds, M.S.J.; Ley, S.V.; Katz, R.B. Physiol. Entomol. 1987, 12, 281-91.
105. Kubo, I.; Ganjian, I. Experientia 1981, 37, 1063-4.

106. Taniguchi, M.; Adachi, T.; Oi, S.; Kimura, A.; Katsumura, S.; Isoe, S.; Kubo, I. Agr. Biol. Chem. 1984, 48, 73-8.
107. Caprioli, V.; Cimino, G.; Colle, R.; Cavagnin, M.; Sodano, G.; Spinella, A. J. Nat. Prod. 1987, 30, 146-51.
108. Cimino, G.; De Ross, S.; De Stefano, S.; Sodano, G. Experientia 1985, 41, 1335-8.
109. Kubo, I.; Miura, I.; Pettei, M.J.; Lee, Y.W.; Pilkiewicz, F.; Nakanishi, K. Tetrahedron Lett. 1977, 4553-6.
110. Hollinshead, D.M.; Howell, S.C.; Ley, S.V.; Mahon, M.; Ratcliffe, N.M. J. Chem. Soc., Perkin I Trans. 1983, 1579-89.
111. Guillerm, D.; Delarue, M.; Jalali-Naini, P; Lallemand, J.Y. Tetrahedron Lett. 1984, 25, 1043-6.
112. Mori, K.; Watanabe, H. Tetrahedron 1986, 42, 273-81.
113. Anonymous. Japanese Patent 58 131 990, 1983.
114. Razmilio, I.; Sierra, J.; Lopez, J.; Cortes, M. Chem. Lett. (Tokyo) 1985, 1113-14.
115. El-Feraly, F.S.; McPhail, A.T.; Onan, K.D. J. Chem. Soc., Chem. Commun. 1978, 75-6.
116. Van Latum, E. Ecoscript 1985, 31, 66 pp.
117. De Groot, A.; Van Beek, T.A. Rec. Trav. Chim. Pays-Bas 1987, 106, 1-18.

RECEIVED November 15, 1988

Chapter 2

North American Ethnobotanicals as Sources of Novel Plant-Based Insecticides

May R. Berenbaum

Department of Entomology, University of Illinois, Urbana, IL 61801–3795

In search for natural products with insecticidal properties to supplant or replace synthetic organic pesticides, the native North American flora has been largely overlooked. Compelling operational reasons for considering North American native plants as sources for insect control chemicals include accessibility and preadaptation of these plants for cultivation in North America. One potential and currently unexploited index to biological activity among the 20,000 species native to North America is the ethnobotanical literature of Native Americans. A compilation of native North American plants described in ethnobotanical accounts as anthelmintic (toxic to "worms"), encompassing publications spanning 1596 to 1980, is presented. Preliminary screenings of extracts of several of purportedly anthelmintic native plants against *Aedes aegypti* revealed insecticidal activity that in some cases corresponded to the presence of known active principles and in other cases may represent the presence of as-yet uncharacterized but potentially useful insecticidal compounds.

"When the plants, who were friendly to man, heard what had been done by the animals, they determined to defeat their evil designs....(The) plants, every one of which has its use if we only knew it, furnish the antidote to counteract the evil wrought by the revengeful animals."
From the Sacred Formulas of the Cherokees, cited in Whitebread, 1934.

There are few attractive alternatives to synthetic organic insecticides for controlling insect pests; however, the use of increasingly higher concentrations of existing insecticides poses a substantial risk to the environment in the form of biomagnification and toxicity to nontarget organisms and the use of alternative structural types is often rendered ineffective by cross or multiple resistance. While only 60 agricultural pests were known to manifest resistance

0097–6156/89/0387–0011$06.00/0

to at least one type of insecticide in 1961, over twenty years later, more than 250 species were known to manifest resistance in one form or another (1). Moreover, regulatory requirements and manufacturing costs, due at least in part to fluctuations in petrochemical supplies, have become so unpredictable that in many cases production and marketing of new pesticides is economically unfeasible for the producer and application unjustifiable for the farmer, based on crop markets and commodity prices. While 19 major pesticides were introduced in the period from 1961 to 1970, only 8 were introduced in the decade from 1971 to 1980 and only 3 in the first half of the period from 1981 to 1990 (2). The rate of introduction of new pesticides is expected to continue to diminish unless major innovations are introduced into development program.

One area of investigation that is not so much an innovation as it is a return to an old approach with new technology is the characterization of secondary plant products. Plants in general produce an enormous variety of chemicals which serve no apparent function in fundamental physiological or biochemical processes; these secondary chemicals (or allelochemicals) are thought to be important in mediating interactions between plants and their biotic environment. Inasmuch as insects are very much a part of the biotic environment of most angiosperm plants, they have no doubt been influential in the evolution of many of these chemicals. That insects are in turn influenced by plant chemicals is evidenced by the successful isolation, characterization, and development of such allelochemicals as nicotine, pyrethrins, and rotenoids as commercial insecticides (3). Even the synthetic organic carbamates are structurally derived from a naturally occurring insecticidal carbamate, physostigmine, produced by Physostigma venenosa (4).

The current resurgence of interest in plant-derived insecticides has focussed on tropical floras. There has long prevailed an assumption, implicit or explicit, that tropical floras are more promising from the point of view of discovering and developing new insecticides (e.g., p. 227, Ref. 3). It is undeniably true that plant taxonomic, and undoubtedly phytochemical, diversity is higher in tropical floras; however the assumptions that phytochemical constituents are accordingly more toxic, due presumably to more intense selection pressure from insect herbivores, or, more importantly, that these compounds are more appropriate for commercial development, do not necessarily follow logically. Attempts to quantify ecogeographic patterns in plant defense have met with less than resounding success (e.g., 5-7). Taxonomic diversity is a relative term. Even though it is less speciose, the North American flora has been largely ignored phytochemically, despite the fact that some of the most spectacular advances in chemical pest control are attributable to North American species (e.g., nicotine from tobacco). There are compelling operational reasons for a closer inspection of the native North American flora as well. Among other things, biologically active plants from North America are preadapted for cultivation in North America; they are less likely to acquire serious new pest problems than are plants grown outside their area of indigeneity; and there is less risk of geopolitical disturbance cutting off supplies, as has historically been the case (e.g., pyrethrum during World War II--see Ref. 3). Even though temperate floras are considerably less speciose than are tropical floras, in North America investigators must still contend with over 20,000 species as potential sources of biologically active chemical constituents. Mass screenings are fraught with operational problems; activity may depend on plant part, manner and time of collection, the choice of bioassay species, and the solvents used for extraction. Even demonstrably toxic species may escape detection in a mass screening. In addition, plant populations are chemically variable and inadequate sampling of a species over its range may create an illusion of inactivity.

Interpreting the North American Ethnobotanical Literature.

One potential (and currently unexploited) solution to the problems of mass screening is to consult the ethnobotanical literature of the region of interest--in this case, the ethnobotany of Native Americans. Folk medicines have traditionally provided insights into promising sources for biologically active materials, particularly medicines, worldwide. Despite the presence of considerable documentation, Native American literature has not been scrutinized carefully to date. Investigators interested in the use of plant extractives or powders for crop protection are likely to be disappointed, since so many Native American cultures were not sedentary agriculturalists. However, considerable use was made of plant materials for protection against ectoparasites; Secoy and Smith (8) cite almost 50 species used to repel or kill fleas, flies, lice, or wound maggots.

One other potential source for information to be used to identify insecticidal plants is to consider those species used as anthelmintics or vermicides. Although various ills have been ascribed to "worms" in a very loose taxonomic sense (9), the majority are recognizably cestodes, nematodes, or possibly dipterans (as in intestinal myiasis). Undoubtedly, a significant number of purportedly anthelmintic plants lack any pharmacological activity and may have been used as medicines due to imagined resemblances between various plant parts and the worms themselves. However, the number of plants said to be anthelmintic is far from a random sample of the North American flora. In a compilation of North American ethnobotanical literature, Moerman (10) reported only 35 out of 1288 species to be anthelmintic. In contrast, 133 species are specifically described as cathartics (an activity potentially easily confused with anthelmintic activity); 577 were described less definitively as "dermatological aids."

Appendix I presents a compilation of native North American plants described in ethnobotanical accounts as anthelmintic. This compilation represents information contained in over 30 herbals, pharmacopoeias, and journal articles covering the period from 1596 to 1980. For older references, plants named in fewer than three sources were not included in the list, due to difficulties in establishing both plant identity and disease diagnosis. Although several species of ferns and even algae appear in several references, only spermatophytes are included in this compilation. A total of 56 species in 45 families were found; rarely was a family represented by more than one or two species or a genus by more than one species.

Screening purportedly anthelmintic plants offers several advantages in winnowing through the North American flora. First, most biologically active plant constituents target basic physiological processes and as such are likely to show considerable activity across invertebrate taxa. In fact, the anthelmintic properties of Nicotiana tabacum were recognized before its insecticidal properties. In 1596, Nicolas Monardes, a Spanish physician and explorer, published the first account of tobacco in his book, "Joyfull Newes out of the New Founde Worlde," and in his account stated that, "...of all Kinds of (worms), it killeth and expelleth them marvellously; the seething of the hearbe made into a Syrope delicately being taken in very little quantitie and the juyce thereof put of the navel" (12). Second, vermicides or vermifuges are almost invariably consumed internally (Nicotiana notwithstanding) and thus acutely toxic materials are not likely to be represented without sufficient warning in terms of administration or dosage (e.g., decoctions of Nicotiana attenuata are to be "taken sparingly,"--Ref. 10-- consistent with the acute toxicity of the presumed active constituent nicotine). Finally, in the event that activity is present, the method of preparation can provide information as to which solvents will be optimal (e.g., extraction in hot vs cold water, alcohol, ingestion of dry powder, etc.).

Bioassays of Anthelmintic Plants for Insecticidal Activity.

In a preliminary feasibility study, I obtained samples from 15 species of native North American plants; 12 species were specifically described as anthelmintic and three species were confamilials without any history of use as anthelmintics (in order to test for effects of bioassay procedures). These species were selected as a subset of suitable test plants due to their availability. Since in the majority of cases anthelmintic preparations are decoctions, that is, extracts prepared in boiling water, all plant material was extracted for 2 h in 50 ml distilled water at 100°C. Different amounts of plant material were used to control for the differences in relative dry weights of plant parts. For seeds, 0.5 g was used; for roots, 1.0 g was used; and for leaves, 2.0 g was used. The insect species used for bioassay was Aedes aegypti (Rockefeller strain) (Diptera: Culicidae). Two replicates of ten neonate larvae were each placed in 20 ml of the plant extract for each species; larvae were provided ad libitum with Tetramin® fish food ground with sand and maintained at 25°C with a 16:8 light:dark photoperiod. Simultaneous distilled water controls were run with all treatments. Survival and development were monitored daily. The proportions eclosing as adults were compared to distilled water controls by G-test and developmental times were compared by Student t-test.

Of the 15 species tested, five proved acutely toxic--complete mortality ensued within five days (Table Ia). Often, first instar

Table Ia. Plant extracts acutely toxic to mosquito larvae (means ± S.E.)

Plant species	Plant part	5-day survival(%)
Angelica atropurpurea	seed	0.0 ± 0
Podophyllum peltatum	root	0.0 ± 0
Sanguinaria canadensis	root	0.0 ± 0
Silene virginica	root	0.0 ± 0
Tephrosia virginiana	seed	0.0 ± 0

Table Ib. Plant extracts inhibiting development of mosquito larvae (means ± S.E.)

Plant species	Plant part	% emerging	Days to emergence
Amorpha canescens	seed	10 ± 0[a]	15.5 ± 0.5[b]
Asclepias incarnata	seed	65 ± 5	12.4 ± 1.9
Cicuta maculata	seed	80 ± 0	16.1 ± 0.3[b]
Euphorbia corollata	seed	15 ± 5[a]	12.0 ± 0.6
Eupatorium perfoliatum	seed	40 ± 10[a]	16.0 ± 0.0[b]
Monarda fistulosa	seed	30 ± 30[a]	11.0 ± 1.2
Robinia pseudoacacia	leaf	15 ± 5[a]	13.7 ± 0.3
Verbena hastata	seed	65 ± 5	12.6 ± 0.8[b]
Distilled water		85 ± 5	8.2 ± 1.0

[a]Significantly different from control ($p < .05$, G-test).

[b]Significantly different from control ($p < .05$, t-test).

Table Ic. Plant extracts without effects on mosquito larvae (means ± S.E.)

Plant species	Plant part	% emerging	Days to emergence
Asarum canadense	root	95 ± 5	8.8 ± 0.4
Betula sp.	leaf	95 ± 5	10.8 ± 0.4
Distilled water		85 ± 5	8.2 ± 1.0

larvae died within 24 hours. Of the remaining extracts, seven caused a significant reduction in survival or increase in development time (Table Ib). Of the three plant species not known to be anthelmintic, only one, Robinia pseudoacacia, significantly affected survival (Table Ic); Asarum canadense and Bedula sp. did not differ significantly from controls.

This assay identified plants in which highly active principles are already known. The two legumes Tephrosia virginica and Amorpha canescens produce rotenoids (3, 8, 13). Angelica atropurpurea contains furanocoumarins (14), plant chemicals already demonstrated to be toxic to mosquito larvae (15, 16). The essential oil of Monarda fistulosa, a mint, contains thymol (17), a highly active ascaricide (and the drug of choice for decades in treatment of hookworm). That this rapid and simple screening yielded several species with known efficacious chemistry indicates that other active plant extracts (e.g., from Silene virginica) may provide additional active chemical constituents. In a second series of bioassays, known chemical constituents of larvicidal or anthelmintic plants (e.g., thymol from Monarda fistulosa, podophyllotoxin from Podophyllum peltatum) were administered to Aedes aegypti neonate larvae as described previously. The activity of several of these chemicals (Table II) was sufficiently high to indicate that commercial use may be economically feasible and competitive with synthetic organic insecticides (15).

Potential Chemicals for Investigation.

Searching for phytochemical similarities in such a diverse assortment of plant taxa is an enterprise almost doomed to failure. Among other things, there is no reason to suppose that plant taxa have converged on similar biochemical defense mechanisms. Indeed, of those plant species for which biologically active materials have been isolated, there is little commonality. Nonetheless, there may be possible biochemical links among these disparate families. In particular, a striking number of purportedly anthelmintic plants are known to produce (or are confamilial with species known to produce) lignans (Table III). Lignans are phenylpropanoid dimers generally linked by the central carbons of their 3-carbon side chains, although in the case of neolignans other types of linkage can form (18). Lignans possess an extraordinary variety of biological properties, including antimitotic, antitumor, and antiviral activity, inhibition of mitochondrial electron transport and energy transfer, inhibition of cAMP photsphodiesterase activity, and cardiovascular disruption. Lignans have not been examined extensively, however, for insecticidal activity. Only a handful of studies (19, 20, 21, 22, 23) have been published, documenting feeding and growth inhibition in several taxa. Several studies have implicated nordihydroguaiaretic acid as a resistance factor against insect damage in Larrea (C. Wisdom, personal communication). Podophyllotoxin, a lignan produced by the anthelmintic and larvicidal plant Podophyllum peltatum, effected complete toxicity to Aedes aegypti in bioassays at concentrations of 5 ppm (Table II). In view of this efficacy, analogues (of podophyllotoxin) and other lignans may well be worth examining more closely for their potential use as insecticides. Although two arylnaphthalene derivatives have been reported to be piscicidal (at levels equivalent to rotenone) (18), there is little *a priori* evidence that these compounds pose a substantial threat to nontarget organisms. Extensive testing, of course, remains to be done. Candidate compounds include sesamin and its derivatives from *Artemisia* (24), or wormwood, species of which are recognized worldwide for anthelmintic properties (25); asarinin, from *Aristolochia*; syringaresinol and lirioresinol from *Liriodendron*; arctiin from the insecticidal *Eupatorium*; and alpha and beta peltatin from *Podophyllum* (17).

Even if lignans prove to be insecticidal to mosquito larvae or other bioassay insect species, they may not necessarily be responsible for the observed insecticidal effects of the plants that produce them. A more appropriate way to identify active agents is to undertake a series of fractionation and purification steps, tracing activity in various fractions via bioassay. However, purification is problematical in that several compounds in combination may be responsible for the overall insecticidal efficacy of the plants. That such may be the case is suggested by the fact that many of the lignans in these plants contain methylenedioxyphenyl (MDP) substituents; compounds of this structure are known to interfere with insect detoxication via cytochrome P450 monooxygenases (26, 27, 28). Since cytochrome P450 monoxygenases are responsible for a wide variety of detoxification reactions, these lignans may serve not only as toxins, by, for example, interfering with cytochrome P450 mediate hormone metabolism (22), but also as synergists for co-occurring toxins. Indeed, another striking characteristic of the plant species in Appendix 1 is the abundance of MDP-containing allelochemicals in the genera to which they belong (Table IV); these compounds include benzylisoquinoline alkaloids, aristolochic acids, and simple phenylpropanoids. The cytochrome P450 inhibiting ability of several of these and related groups has been documented (benzylisoquinoline alkaloids-29; simple phenylpropanoids-27, amides-30; protopine alkaloids and dibenzylbutyrolactones-31.

Chinese herbalists have long recognized the value of mixtures as curatives; prescriptions, for example, often include "monarch" (jun), "his subject" (chen), "assistant" (zuo), and "messenger" (shi) ingredients (32). Purification of single active components, the usual goal in screening studies, may yield a less than optimal product, particularly if MDP-mediated inhibition of detoxification is involved. Plant extracts may be more efficacious than purified derivatives due to synergistic or potentiating interactions among plant constituents--compounds, for example, which facilitate penetration or delivery of the toxin to the target site working in combination with others that, as MDP-containing compounds, interfere with the ability of an organism to detoxify or excrete a principal toxin. According to Vogel (33), "While Indians sometimes compounded simples, they seldom used more than two or three ingredients"--it may well be that combining different plants increases the opportunities for antagonistic, rather than synergistic, interaction among plant constituents. Co-occurring synergists may be structurally optimized for intraspecific phytochemistry (27).

Table II. Toxicity of active constituents of larvicidal or anthelmintic plants to mosquito larvae (means ± S.E.)

Compound	Concentration	3-day survival (%)
Aristolochic acid	1 ppm	0.0 ± 0.0
Podophyllotoxin	5 ppm	0.0 ± 0.0
Thymol	10 ppm	0.0 ± 0.0

Table III. North American plant families reported to contain lignans (lignan information from MacRae and Towers 1984)

Aristolochiaceae*	Myrtaceae
Anacardiaceae*	Oleaceae
Apocynaceae*	Phytolaccaceae
Betulaceae*	Polygalaceae
Berberidaceae*	Rosaceae*
Bignoniaceae*	Rutaceae*
Compositae (Asteraceae)*	Salicaceae*
Convolvulaceae	Saururaceae
Cucurbitaceae*	Schisandraceae
Ericaceae	Scrophulariaceae*
Euphorbiaceae*	Simaroubaceae
Fagaceae*	Solanaceae*
Lauraceae*	Styracaceae
Leguminosae (Fabaceae)*	Ulmaceae*
Linaceae	Urticaceae
Loranthaceae	Umbelliferae (Apiaceae)*
Magnoliaceae*	Verbenaceae*

*Families with anthelmintic plants

Table IV. Purported anthelmintic genera with methylenedioxyphenyl substituents (from 17 and 34)

Angelica	*Eupatorium*
Aristolochia	*Populus*
Artemisia	*Podophyllum*
Cirsium	*Sanguinaria*
Liriodendron	

The Cherokee Indians believed that human ills result from a compact on the part of all the animals who felt themselves injured by humans--from the bears and deer who were hunted for food down to the insects and smaller animals, "trodden upon without mercy, out of pure carelessness or contempt" (9). According to the legend, plants, with no grudges to bear, agreed to furnish remedies for some of the ills to which humans would be subjected. It is indeed true that humans, through carelessness or contempt, have exacted enormous inadvertent mortality upon nontarget organisms through ineffective or inattentive use of synthetic organic insecticides; it may be possible that plants can provide the wherewithal to alleviate the mortality in the form of specific biodegradable alternatives to synthetic organic insecticides.

Appendix I--Native North American plants that are purportedly anthelmintic (order of plant families according to 35)

Pinaceae
- *Abies fraseri*

Cupressaceae
- *Juniperus virginiana*, *J. osteosperma*
- *Thuja occidentalis*

Typhaceae
- *Typha* spp.

Araceae
- *Symplocarpus foeditus*

Liliaceae
- *Aletris farinosa*
- *Chamaelirium luteum*

Salicaceae
- *Populus tremuloides*
- *Salix* spp.

Myricaceae
- *Myrica cerifera*

Juglandaceae
- *Juglans cinerea*

Betulaceae
- *Corylus rostrata*, *C. cornuta*

Ulmaceae
- *Ulmus rubra*

Moraceae
- *Morus rubra*

Aristolochiaceae
- *Aristolocia serpentaria*

Chenopodiaceae
- *Chenopodium ambrosioides var. anthelminticum*

Nyctaginaceae
- *Mirabilis nyctaginea*

Caryophyllaceae
- *Silene virginica*

Berberidaceae
- *Podophyllum peltatum*

Magnoliaceae
- *Liriodendron tulipifera*

Annonaceae
    Asimina triloba
Lauraceae
    Lindera benzoin
Papaveraceae
    Sanguinaria canadensis
Saxifragaceae
    Ribes americanum
Rosaceae
    Prunus americana, *P. serotina*
Leguminosae
    Amorpha canescens
    Tephrosia virginiana
Rutaceae
    Ptelea trifoliata
Ebenaceae
    Diospyros virginiana
Euphorbiaceae
    Euphorbia corollata
Anacardiaceae
    Rhus typhina
Aquifoliaceae
    Ilex verticillatus
Aceraceae
    Acer spicatum
Nyssaceae
    Nyssa sylvatica
Umbelliferae
    Angelica atropurpurea
    Eryngium aquaticum
Loganiaceae
    Spigelia marilandica
Gentianaceae
    Sabatia angularis
Apocynaceae
    Apocynum cannabinum
Asclepiadaceae
    Asclepias incarnata
Verbenaceae
    Verbena hastata
Labiatae
    Monarda fistulosa
Solanaceae
    Nicotiana tabacum, *N. attenuata*
Caprifoliaceae
    Lonicera dioica
Valerianaceae
    Valeriana ciliata
Scrophulariaceae
    Chelone glabra
Bignoniaceae
    Catalpa bignonioides
Cucurbitaceae
    Cucurbita pepo

Campanulaceae
- *Lobelia cardinalis*

Compositae
- *Cirsium pulchellum*
- *Eupatorium perforatum*

Chronological Bibliography to Appendix I

N. Monardes, Joyfull News Out of the New Founde Worlde, London, 1596, "Englished by J. Frampton."
M. Cutler, An Account of Some of the Vegetable Productions Naturally Growing in this Part of America (New England), Philadelphia: Philosophical Society, 1785.
J. D. Schoepf, Materic Medica Americana Potissimum regni Vegetabilis, Erlangae, 1787.
S. Stearns, An American Herbal or Materia Medica, Carlisle, Thomas and Thomas, 1801.
W. Meyrick, The New Family Herbal, Birmingham, 1802.
B. S. Barton, Collections for an Essay Towards a Materia Medica of the United States, Philadelphia, 1804.
P. Smith, The Indian Doctor's Dispensatory, Cincinnati: Browne and Looker, 1812.
S. Henry, American Family Herbal, New York, 1814.
W. Barton, Vegetable Materia Medica of the United States, Philadelphia: Carey and Sons, 1817.
T. Green, Universal Herbal, containing an account of all the known plants in the world, London: Caxton Press, 1824.
J. Bigelow, American Medical Botany, Boston: Hilliard and Melcalf, 1817-20.
P. Bowker, The Indian Vegetable Family Instructor, Boston: 1836.
P. Good, The Family Flora and Materia Medica Botanica, Elizabethtown, 1845.
R. Griffith, Medical Botany, Philadelphia: Lea and Blanchard, 1847.
A. Clapp, A Synopsis or Systematic Catalogue of the Medicinal Plants of the United States, Philadelphia: T. K. and P. G. Collins, 1852.
W. Darlington, American Weeds and Useful Plants, New York: A. O. Moore, 1859.
O. Brown, The Complete Herbalist, Jersey City, 1865.
L. Johnson, A Manual of the Medical Botany of North America, New York: William Wood and Co., 1884.
C. Millspaugh, American Medicinal Plants, Philadelphia: Boericke and Tafel, 1887.
W. Fernie, Herbal Simples, Bristol: John Wright and Co., 1895.
A. Henkel, Weeds Used in Medicine, Farmers' Bulletin 188, USDA, Washington: USDA, 1906.
W. Stockberger, The Drug Known as Pink Root, Bulletin 100, Washington: USDA, 1906.
A. Henkel, American Root Drugs, Bulletin 107, Washington: USDA, 1907.
A. Henkel, American Medicinal Flowers, Fruits, and Seeds, Bulletin 26, Washington: USDA, 1913.

E. Stuhr, Manual of Pacific Coast Drug Plants, Lancaster: Science Press Printing Co., 1933.
M. Grieve, A Modern Herbal, New York: Hafner Publishing Co., 1967 (Originally published in 1931).
L. Curtin, Healing Herbs of the Upper Rio Grande, Laboratory of Anthropology, Santa Fe, 1947.
E. Steinmetz, Materia Medica Vegetabilis, Amsterdam, 1954.
A. Krochmal, A Guide to the Medicinal Plants of Appalachia, USDA Forest Service Research Paper NE-138, 1969.
V. Scully, A Tresury of American Indian Herbs, New York: Crown Publishers, Inc., 1970.
M. Weiner, Earth Food-Earth Medicine, New York: MacMillan Co., 1972.
A. Krochmal, Guide to the Medicinal Plants of the United States, New York, 1973.
D. Moerman, 1977. American Medical Ethnobotany: A Reference Dictionary. NY: Garland Pub., Inc.
Angier, B., 1978. Field Guide to Medicinal Wild Plants. Harrisburg (PA): Stackpole Books.

Herbals without dates:

R. Brook, A New Family Herbal, Huddersfield,
M. Robinson, The New Family Herbal, London: W. Nicholson and Son,

In plant identification and in determining nonmenclature, the following were consulted:

M. Fernald, Gray's Manual of Botany, 8th ed., New York: American Book Co., 1950.
H. Kelsey and W. Dayton, ed., Standardized Plant Names, Harrisburg: J. Horace MacFarland, 1942.
G. Lawrence, Taxonomy of Vascular Plants, New York: The MacMillan Co., 1951.
Kartesz, J.T. and R. Kartesz, 1980. A synonymized checklist of the vascular flora of the United States, Canada and Greenland. Chapel Hill: University of North Carolina Press.

Acknowledgments:

I thank A. Zangerl for assistance with statistics and unflagging interest, R. Novak for eggs of *Aedes aegypti*, and J. Parrish for assistance in obtaining plant material. This research was supported in part by N.S.F. BSR83-1407.

Literature Cited:

1. Georghiou, G. and R.B. Mellon, 1983. Pesticide resistance in time and space. Pages 1-46 in G.P. Georghiou and T. Saito, eds. Pest Resistance to Pesticides. NY: Plenum Pub.
2. Ku, H.S., 1987. Potential industrial applications of allelochemicals and their problems. A.C.S. Symp. Ser. 330: 449-454.
3. Jacobson, M. and D.G. Crosby, 1971. Naturally Occurring Insecticides. NY: Marcel Dekker Inc.

4. Balandrin, M., J. Klocke, E. S. Wurtele, and W. H. Bollinger, 1985. Natural plant chemicals: sources of industrial and medicinal materials. Science 228: 1154-1160.
5. Levin, D., 1976. Alkaloid-bearing plants; an ecogeographic perspective. Am. Nat. 110: 261-284.
6. Levin, D., and B. York, 1978. The toxicity of plant alkaloids: an ecogeographic perspective. Biochem. Syst. Ecol. 6: 61-76.
7. Moody, S., 1976. Latitude, continental drift and the percentage of alkaloid-bearing plants in floras. Am. Nat. 110: 965-968.
8. Secoy, D.M. and A. E. Smith, 1983. Use of plants in control of agricultural and domestic pests. Econ. Bot. 37: 28-57.
9. Whitebread, C., 1934. The Indian medical exhibit of the Division of Medicine in the United States National Museum. Proc. U.S.N.M. 67 (10): 1-26.
10. Moerman, D., 1977. American Medical Ethnobotany: A Reference Dictionary. NY: Garland Pub., Inc.
11. Berenbaum, M.R., 1988. Allelochemicals in insect-microbe-plant interactions: agents provacateurs in the coevolutionary arms race. Pages 97-123 in Novel Aspects of Insect-Plant Interactions (P. Barbosa and D. Letourneau, eds.). NY: J. Wiley and Sons.
12. Monardes, N., 1596. Joyfull Newes out of the New-Founde Worlde. London, "Englished by J. Frampton."
13. Jacobson, M., 1982. Plants, insects and man--their interrelationships. Econ. Bot. 36: 346-354.
14. Berenbaum, M., 1981. Furanocoumarin distribution and insect herbivory in the Umbelliferae: plant chemistry and community structure. Ecol. 62: 1254-1266.
15. Wat, C.-K., S.K. Prasad, E.A. Graham, S. Partington, T. Arnason, and G.H.N. Towers, 1981. Photosensitization of invertebrates by natural polyacetylenes. Biochem. Syst. Ecol. 9: 59-62.
16. Kagan, J., P. Szczepanski, V. Bindokas, W. Wulff, and J. S. McCallum, 1986. Delayed phototoxic effects of 8-methoxypsoralen, khellin, and sphondin in Aedes aegypti. J. Chem. Ecol. 12: 899-914.
17. Hegnauer, R., 1969-1973. Chemotaxonomie der Pflanzen. Basel: Birkhauser Verlag.
18. MacRae, W.D. and G.H.N. Towers, 1984. Biological activities of lignans. Phytochem. 23: 1207-1220.
19. Russel, G.B., P.Singh, and P. Fenmore, 1976. Insect-control chhemicals from plants III. Toxic lignans from Libocedrus bidwilli. Aust. J. Biol. Sci. 29: 99-103.
20. Kamikado, T., C.F. Chang, S. Murakoshi, A. Sakurai, and S. Tamura, 1975. Agric. Biol. Chem. 39: 833- .
21. Isogai, A., S. Murakoshi, A. Suzuki, and S. Tamura, 1973. Structures of new dimeric phenylpropanoids from Myristica fragrats Houtt. Agric. Biol. Chem. 37: 1479-1486.
22. Bowers, W., 1968. Juvenile hormone: activity of natural and synthetic synergists. Science 161: 895-97.
23. Wada, K. and K. Munakata, 1970. (-) Parabenzlactone, a new piperolignanolide isolated from Parabenzoin trilobum Nakai. Tetrahedron Let. 23: 2017-2019.

24. Gregor, H., 1979. Polyacetylene und Sesamine als chemische Merkmale in der Artemisia absinthium Gruppe. Planta Medica 35: 84-91.
25. Berenbaum, M., 1975. Anthelmintic plants in American botanic medicine. Yale Scientific (Summer 1975): 9-16.
26. Berenbaum, M., 1985. Brementown revisited: allelochemical interactions in plants. Rec. Adv. Phytochem. 19: 139-169.
27. Berenbaum, M. and J. Neal, 1985. Synergism between myristicin and xanthotoxin, a naturally co-occurring plant toxicant. J. Chem. Ecol. 11: 1349-1358.
28. Berenbaum, M. and J. Neal, 1987. Allelochemical interactions in crop plants. A.C.S. Symp. Ser. 330: 416-430.
29. Dalvi, R., 1985. Sanguinarine: its potential as a liver toxic alkaloid present in the seeds of Argemone mexicana. Experientia 41: 77-78.
30. Miyakado, M., I. Nakayama, N. Ohno, and H. Yashioka, 1983. Structure chemistry, and actions of the Piperaceae amides: new insecticidae constituentsd isolated from the pepper plant. In Natural Products for Innovative Pest Management, D.L. Whitehead, ed. NY: Pergamon Press, pp. 369-382.
31. Neal, J., 1987. Ecological aspects of insect detoxication enzymes and their interaction with plant allelochemicals. Doctoral dissertation, University of Illinois at Urbana-Champaign.
32. Ren-Sheng, X., Z. Quiao-Zhen, and X. Yu-Yuan, 1985. Recent advances in studies on Chinese medicinal herbs with physiological activity. J. Ethnopharmacol. 14: 223-253.
33. Vogel, V. J., 1970. American Indian Medicine. Norman: University of Oklahoma Press.
34. Newman, A.A., 1962. The occurrence, genesis and chemistry of the phenolic methylene-dioxy ring in nature. Chem. Prod. 1962 (March): 115-118, 161-171.
35. Lawrence, G.H.M., 1951. Taxonomy of vascular plants. NY: The MacMillan Company.

RECEIVED November 18, 1988

Chapter 3

Search for New Pesticides from Higher Plants

A. Alkofahi[1,4], J. K. Rupprecht[1], J. E. Anderson[1], J. L. McLaughlin[1], K. L. Mikolajczak[2], and Bernard A. Scott[3]

[1]Department of Medicinal Chemistry and Pharmacognosy, School of Pharmacy and Pharmacal Sciences, Purdue University, West Lafayette, IN 47907

[2]Northern Regional Research Center, Agricultural Research Service, U.S. Department of Agriculture, Peoria, IL 61604

[3]Lilly Research Laboratories, Greenfield Laboratories, P.O. Box 708, Greenfield, IN 46140

Pest control methods of the near future will include potent, more selective, and bio-degradable pesticides discovered as the natural protectants of higher plants. A number of botanical pesticides (pyrethrins, rotenoids, nicotine, the natural isobutylamides, quassia, sabadilla, hellebore, and ryania) are enjoying expanding commercial uses. Some chemical companies are developing additional plant extracts from the Meliaceae and Rutaceae containing potent limonoids such as azadirachtin. Our laboratories have been involved in pesticidal screening programs of higher plant extracts and have already screened several hundred plant species against a battery of indicator pests. Lists of promising leads have been tabulated, and bioactivity-directed fractionations, using brine shrimp lethality as a simple in-house bioassay, are in progress. Our first product is asimicin, a novel acetogenin (polyketide) from the bark of the paw paw tree, *Asimina triloba* (Annonaceae); this group of very potent compounds now offers a host of chemical and biological challenges for further research and commercial exploitation.

The demand for pesticidal and herbicidal compounds to control animal and plant pests and weeds has created an agro-chemical business in the U.S. which totals over $15 billion annually. These synthetic wonders have facilitated astonishing gains in agricultural production, but these same compounds, in some cases,

[4]Current address: Faculty of Pharmacy, Jordan University of Science and Technology, Irbid, Jordan

0097–6156/89/0387–0025$06.00/0

have posed serious problems to health and environmental safety. Newer, more selective, and biodegradable compounds must replace these generally toxic and persistent chemicals of the present and the immediate past. Newer pest control methods will include: a) use of natural predators, parasites, and pathogens; b) breeding resistant varieties of crop species; c) pest sterilization techniques; d) use of mating and feeding attractants in combination with traps; e) development of interfering hormones; f) improved methods for grain storage; and g) discovery and development of better, more specific, and biodegradable, pesticides (1,2).

The expanding use of synthetic pyrethroids, which are based on natural prototypes from chrysanthemum flowers, is a good example of the desired approach to better pesticides (3); these compounds are more specific for killing insects and exhibit few negative effects on plants, livestock, or humans; in addition, there is often less resistance developed to these compounds in the target insects. A number of additional botanical insecticides (rotenoids, nicotine, the natural isobutylamides, quassia, sabadilla, hellebore, and ryania) are enjoying expanding commercial uses (4), and a number of chemical companies are evaluating (and in some cases developing) extracts from plants of the Meliaceae and Rutaceae containing insecticidal limonoids such as azadirachtin (5,6).

To protect themselves from being eaten, plants have, in effect, been waging biochemical warfare for thousands of years against insects and herbivores (7,8). Perhaps many plant species which failed to develop such protective "secondary metabolites" were consumed and made extinct; thus, the potential for finding pesticides in higher plants should be high. The job of the phytochemist is to detect, isolate, and identify these compounds. Martin Jacobson has spent much of his research career encouraging an expansion of research in this area (1,4,7,8). The prospect of discovery and exploitation of these novel higher plant metabolites, useful as pesticides or pesticide prototypes, seems excellent and is especially timely. For the past several years, our laboratories have been involved in dedicated programs of bioactivity-directed screening and fractionation of higher plant extracts to yield such agents.

SCREENING HIGHER PLANTS FOR PESTICIDES

The single, most important, factor in the search for new bioactive substances is the convenience and reliability of the bioassay systems. Screening bioassays must be inexpensive and rapid, have broad application to numerous target organisms, be reproducible and statistically valid, and require very little of the test substance. No single pest model meets all of these requirements, and a battery of indicator organisms from various taxa is, undoubtedly, the best approach to screening. However, financial constraints usually force the choice of one or only a few.

EUROPEAN CORN BORER BIOASSAY AND *THEVETIA THEVETIOIDES*

About ten years ago at the Peoria USDA laboratory, a plant screening program for pesticides was initiated using 7-day-old larvae of the European corn borer, *Ostrinia nubilalis* (9). This insect was chosen because of its economic importance to agriculture in the midwest. Seeds of several dozen plant species, from the extensive USDA seed collection in Peoria, were extracted in a Soxhlet sequentially with hexane and ethanol, and the extract residues were incorporated into the corn borer diet. Mortalities were determined, generally after nine days. This screen revealed several species which seemed worthy of fractionation (10).

One of the most promising leads was the ethanol extract of the defatted seeds of *Thevetia thevetioides* (HBK) K. Schum. (Apocynaceae). Large-scale extraction and fractionation of seeds from Mexico was laboriously monitored with the corn borer bioassay. Percent mortalities were determined for partitioned fractions and for pools of similar combined fractions from large chromatography columns. The active pools were rechromatographed and tested with the corn borer bioassay at every step, until two insecticidal components were crystallized. Using spectral (^{1}H nmr, ^{13}C nmr, ir) and physical (mp, tlc) methods, the two compounds were identified as neriifolin and 2'-acetylneriifolin, cardioactive glycosides, previously known in other species of the Apocynaceae (11).

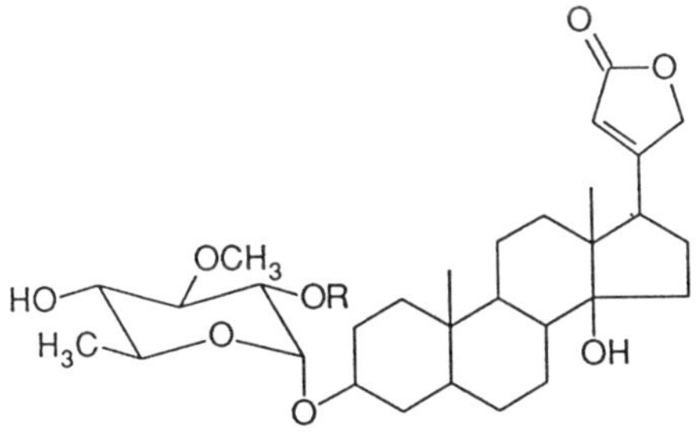

Neriifolin (corn borer LD_{50} 30 ppm in the diet) was ca. six times as active as 2'-acetylneriifolin (LD_{50} 192 ppm) and showed excellent dose response curves with larval mortality. Under such test conditions, carbofuran, a popular commercial insecticide for corn borers, has an LD_{50} of 1-2 ppm. Further testing of neriifolin showed potent insecticidal activities against the codling moth (*Cydia pomonella*), striped cucumber beetle (*Acalymma vittatum*), and the Japanese beetle (*Popilla japonica*). The two compounds were also significantly active in the 9KB (human nasopharynageal carcinoma) cytotoxicity assay (neriifolin ED_{50} 2.2 x 10^{-2} mcg/ml and 2'-acetylneriifolin ED_{50} 3.3 x 10^{-2} mcg/ml) (11). In mammals, these glycosides are cardiotoxic and are quite potent (LD_{50} in cats, 0.196 mg/kg and 0.147 mg/kg, respectively); extrapolation to a 70 kg human would put the fatal dose at about 10-14 mg; as a comparison the fatal dose of hydrogen cyanide for humans is 50-100 mg. However, in human poisoning with cardiac glycosides, as found iin red squill, emesis should prevent fatalities after oral ingestion. In our study, cardiac glycosides were proven to be very potent as natural insecticides, and this protective effect likely explains the widespread distribution of these glycosides in several plant families.

A U.S. patent application was subsequently filed to claim the use of these agents in the control of insect pests (12), but the resulting publication (11) preceded the filing date of the patent application by a few days over one year, and the patent was disallowed. Nonetheless, we remained convinced that these procedures could work to yield new natural pesticides. However, a review of the project suggested a major need for improvement. The corn borer bioassay was too labor-intensive and it rapidly depleted our extracts.

BRINE SHRIMP BIOASSAY

To facilitate continued research for such useful bioactive materials from higher plants, a convenient general bioassay was needed. Specific bioassays, e.g., receptor binding, enzyme inhibition, etc., often overlook other useful activities which are not detected, or are ignored, in the screening process. There is a real need for reliable, general bioassays which can detect the broad spectrum of bioactivities present in higher plant extracts and, yet, can be employed by natural product chemists, in house, at low cost, using a minimum amount of test materials, to permit phytochemical screening and to guide fractionation.

Since most active plant principles are toxic at elevated doses, our approach has been to develop a general bioassay that simply screens for lethality in the simplest of macroscopic zoologic systems. Once the active plant species have been detected, a battery of more expensive and specialized bioassays could then suggest the best candidates for large scale fractionation. Our premise here has been that toxicology is simply pharmacology at a higher dose, or pharmacology is simply toxicology at a lower dose. Thus, a general bioassay for lethality might lead to new useful pharmacologic agents. However, if the agents prove to be persistently toxic, with narrow therapeutic indexes, and useless as drugs, we might then promote them as pesticides.

After considering several possible organisms, we chose brine shrimp (*Artemia salina*), a tiny crustacean, as our general bioassay tool (13). The eggs of brine shrimp are readily available at low cost in pet shops as a food for tropical fish. The eggs remain viable for years, especially if refrigerated, in the dry state. Upon being placed in a brine solution, the eggs hatch within 48 hours and swim toward a light source, providing large numbers of larvae (nauplii). Compounds and extracts are tested initially at concentrations of 10, 100, and 1000 ppm after being placed in vials containing 5 ml of brine and ten shrimp in each of three replicates. Survivors are counted after 24 hours, and the percentage of deaths at each dose is recorded. Intermediate dosages can be tested once the effective dosage range is determined. These data can then be used, in a simple program on an IBM personal computer, to estimate LC_{50} values and 95% confidence iintervals. The LC_{50} values, thus, give us a single value for the comparison of potencies among various extracts. These procedures are summarized in Table I (14).

Table I. Materials and Procedures for Brine Shrimp Lethality Bioassay

---

A. Materials

1. *Artemia salina* Leach (brine shrimp eggs from store)
2. Sea salt (from fish store)
3. Small tank with perforated dividing dam to grow shrimp, cover, and lamp to attract shrimp.
4. Syringes; 5 ml, 0.5 ml, 100 mcl, and 10 mcl.
5. 2 dram vials (9 per sample + 1 control)

B. Procedures

1. Make sea water according to directions on box (ca. 38 g sea salt per liter of water), filter.
2. Put sea water in small tank and add shrimp eggs to one side of the divided tank, cover this side. Lamp on other side will attract shrimp through perforations in dam.
3. Allow 2 days for the shrimp to hatch and mature.
4. Prepare vials for testing; for each fraction, test initially at 1000, 100, and 10 mcg/ml; prepare 3 vials at each concentration for a total of 9 vials; weigh 20 mg of sample and add 2 ml of solvent (20 mg/2 ml); from this solution transfer 500, 50, or 5 mcl to vials corresponding to 1000, 100, or 10 mcg/ml, respectively. Evaporate solvent under nitrogen and then put under high vacuum for about 30 min; volatile solvents will evaporate over night. Alternatively, materials may be dissolved in DMSO, and up to 50 mcl may be added/5 ml brine before DMSO toxicity affects the results.
5. After 2 days (when the shrimp larvae are ready), add 5 ml sea water to each vial and count 10 shrimp per vial (30 shrimp per dilution).
6. 24 hours later count and record the number of survivors.
7. Analyze data with Finney computer program to determine LC_{50} values and 95% confidence intervals. A copy of this program for IBM PC's is available from Dr. McLaughlin.
8. Additional dilutions at less than 10 mcl/ml may be needed for potent materials; intermediate concentrations can be prepared and tested to narrow the confidence intervals.

Since 1982, the brine shrimp bioassay, combined with specific antitumor bioassays, has led us to several in vivo active plant antitumor agents (15, 16 inter alios). In addition, we have used this system in the screening and fractionation of pesticidal plant extracts as described below. It has become the "work horse" bioassay of our laboratory. Each investigator performs his/her own brine shrimp tests and quickly (24 hours) has the results. The results are quantitative and quite reproducible within the 95% confidence limits. Our contention is that toxicity to this simple crustacean is a convenient indicator of toxicity to invertebrate economic pests.

Plants and Extracts Involved in Screening

Our Department of Medicinal Chemistry and Pharmacognosy at Purdue has an established and continuing interest in the chemistry of bioactive natural products. In recent years, we have focused primarily on antitumor compounds isolated and characterized from higher plants through the support of grants and contracts from the National Cancer Institute (NCI). In our warehouse now are varying quantities (0.5 - several kg) of over 2000 species of interesting higher plants. These, hereafter referred to as "warehouse plants" (Table II), for the most part, do not represent random collections but have folkloric or poisonous histories, were lethal to mice in the NCI antitumor screen, were cytotoxic, or have other reasons for us to suspect that they may possess interesting biological activities. This collection represents thousands of dollars already spent in acquisition, shipping, botanical authentication, drying, milling, and storage. In addition, extracts of nearly 400 plant species were on hand from the National Cancer Institute (McCloud Collection) and through a collaborative project with King Saud University in Riyadh, Saudi Arabia (Table II). A continuing flow of new plant materials is maintained through several international collaborators (Iran, China, India, Sri Lanka, Thailand, Panama, Jordan, Venezuela, Brazil).

Screening of all these plant extracts was initiated in a previous contract (1985-86) between Eli Lilly (Greenfield) and Purdue University. Approximately five new warehouse plant species per week were extracted, partitioned and screened. Screening data was obtained on extracts of 739 species. The extracts represent the three different sets of plant materials as summarized in Table II.

Extracts of the warehouse plants and a few of the most active extracts of the Saudi plants were partitioned through Scheme 1 to provide extracts F001-F006. Samples of approximately 100 mg of each initial extract or partitioned fraction were submitted, through weekly mailings at ca. 30/week, to Greenfield. At Lilly, tests are conducted in seven indicator organisms (mosquito larvae, blowfly larvae, *Caenorhabditis elegans* or *Halmonchus contortus*, corn rootworm, two-spotted spider mite, southern army worm, and melon aphid). The extracts of the active McCloud plants were also assayed for brine shrimp lethality in our laboratory. Corn rootworm activity, at 300 ppm, was difficult to detect, and southern army worm (SAW), at even 5000 ppm, was resistant to all but a few species. The free-living nematode, C. elegans, gave too many positives in the first year, and in the second year was replaced by H. contortus which is more selective.

Results of extraction and partitioning (Scheme 1) of Chrysanthemum flowers (containing pyrethrins) and Lonchocarpus roots (containing rotenone), as controls, demonstrated the effectiveness of these procedures at detecting such desirable activities and in enriching the pesticidal materials. As a further control, extracts of *Warburgia salutaris* (*W. ugandensis*) (Canellaceae), screened as positive in the pesticidal assays and, by activity-directed fractionation monitoring with brine shrimp, yielded the expected drimane sesquiterpenes, warburganal and

Plant: ______________ genus ______________ species (______________ family)

Identification No.: ______________

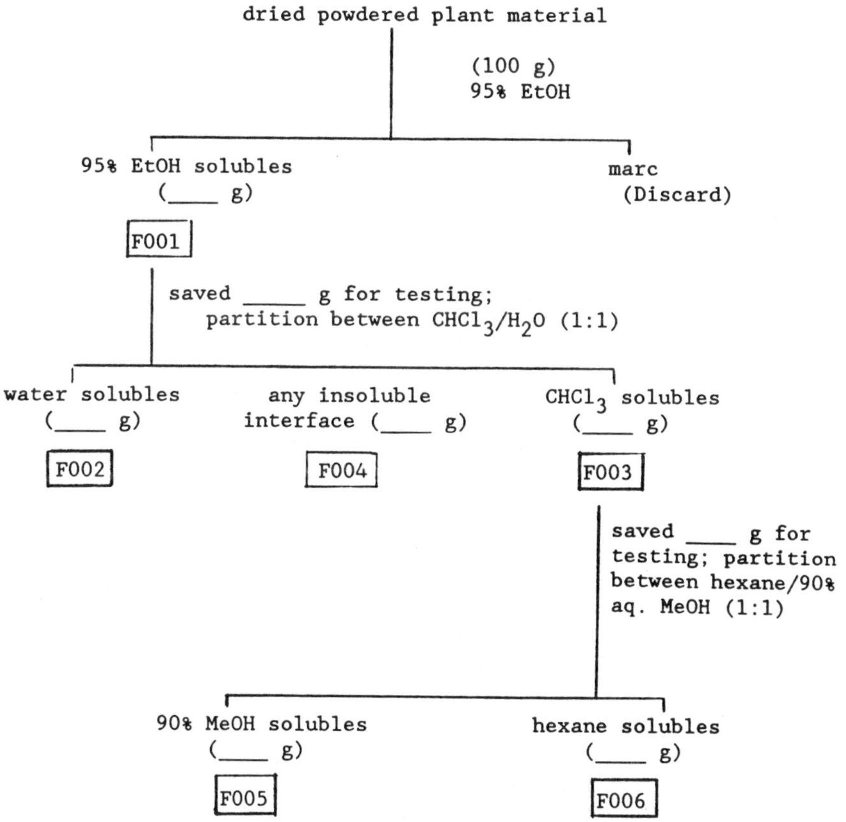

dried powdered plant material

(100 g) 95% EtOH

95% EtOH solubles (____ g) F001

marc (Discard)

F001: saved _____ g for testing; partition between $CHCl_3/H_2O$ (1:1)

water solubles (____ g) F002

any insoluble interface (____ g) F004

$CHCl_3$ solubles (____ g) F003

F003: saved ____ g for testing; partition between hexane/90% aq. MeOH (1:1)

90% MeOH solubles (____ g) F005

hexane solubles (____ g) F006

Submitted fractions (   ) to __________ for __________ testing; Date __________

Scheme 1. Standard Flow Sheet of Extraction and Initial Partitioning to Provide Screening Extracts for Bioassays

Table II. Sources of Plant Materials Screened

A. Warehouse Collection: (EtOH extracts all partitioned to give F001-F006, see Scheme 1) 359 species (98 families)

B. Saudi Collection: (EtOH or $CHCl_3$ extracts; a few actives partitioned to give F001-F006), 136 species (43 families)

C. McCloud Collection: (CH_2Cl_2 extracts; none partitioned), (from NCI) 244 total species: 92 actives (46 families) and 152 inactives (56 families)

muzigadial (17, 18) which are known to exhibit potent insect antifeedant (19-21) and molluscicidal (22) activities.

The pesticidal screening results were tabulated on computers both at Greenfield and in our laboratory at Purdue. The computer programs permit tabulation of extracts and species active against specific pests or against various combination of pests. Choosing those with the broadest spectrum of activities, we selected twenty target plant species for large-scale, bioactivity-directed, fractionation. A literature search shows that, aside from our work, no previous bioactivity-directed research has been performed on these target species. One of the target species, the common paw paw, *Asimina triloba*, Annonaceae, has been a subject of our fractionation efforts for the past five years. This work has now been successful; the bioassay-directed fractionation has not been published previously, and will be discussed below. Proprietary interests prevent us from disclosing additional target species and the screening results.

THE PAW PAW PROJECT

Isolation Procedure

Extraction of Plant Material. Bark of *Asimina triloba* (paw paw), collected at the Purdue Horticulture Farm, was dried and ground in a Wiley mill to 2 mm. The ground bark (3.990 kg) was extracted by exhaustive percolation with 185 l of 95% ethanol. The ethanol solubles were vacuum evaporated to a syrupy residue which was labeled Fraction F017. Fraction F017 was partitioned between CH_2Cl_2/H_2O (1:1), and the water solubles were taken to dryness and labeled F018. The CH_2Cl_2 solubles were recovered as a syrupy residue using a solvent evaporator. This residue was labeled Fraction F019, and a 223 g sample of F019 was then partitioned between hexane and 90% aqueous methanol. The methanol solubles were thereafter vacuum evaporated to a thickened syrup (41 g) and labeled as Fraction F020. The recovered hexane solubles constituted Fraction F021. The results of the brine shrimp bioassay on the fractions are reported in Table III.

Table III. Assay of Partition Fractions from Ethanolic Extract of the Bark of *Asimina triloba*

Fraction No.	Brine shrimp LC_{50} (ppm)
017 (ethanol extract)	7.56
018 (H_2O solubles)	>1000
019 (CH_2Cl_2 solubles)	1.67
020 (aqueous methanol solubles)	0.04
021 (hexane solubles)	715
asimicin	0.03

Chromatography of Fraction F020. A sample of F020 (39.5 g), the most toxic partition fraction indicated by the brine shrimp assay (Table III), was adsorbed on celite (150 g) and applied to a silica column (4.0 kg) packed in benzene/EtOAc (80:20). The column was eluted with 10-liter aliquots of 20%, 50% EtOAc/benzene, 100% EtOAc, 2%, 5%, 20%, 50% MeOH/EtOAc, and finally, 100% MeOH, and fractions were dried and weighed.

A TLC plate was run, in hexane/EtOAc (20:80), on every fifth column fraction and sprayed with 0.5% tetrazolium blue in MeOH:5N NaOH (1:1). From the appearance after TLC, pools of similar compounds were made and the pools assayed with brine shrimp. The results are reported in Table IV.

C-18 Column Chromatography of Fraction AT 49. The most active fraction in the brine shrimp assay, AT 42, (Table IV) was consumed in development of a satisfactory separation method. The next most active and comparable fraction, AT 49, was subjected to C-18 column chromatography; AT 49 (1.924 g) was adsorbed on celite (8 g), applied to the top of a column of C-18 silica (60 g), and chromatographed and assayed for brine shrimp lethality as shown in Table V.

Chromatotron Separation of Fraction AT 49-5. Separation of the most abundant active fraction AT 49-5, which was active in the brine shrimp assay, from the above C-18 column separation of AT 49 (Table V) was carried out as follows. A 4 mm silica chromatotron rotor was loaded with 586.3 mg of AT 49-5 and eluted with $CHCl_3/MeOH/H_2O$ (5:2:2); this resulted in 25 fractions which were then pooled on the basis of their similarities upon TLC analysis (Table VI).

Purification of Fraction AT 49-5-2. A white waxy substance (m.p. 68x-69x C) precipitated from an Et_2O solution at AT 49-5-2 upon the addition of hexane. The precipitate was collected, assayed in five different TLC systems, and visualized with $Na_2Cr_2O_7$ in 40% H_2SO_4. Each of the TLC analyses showed only a single spot

Table IV. Toxicity of Column Chromatography Fractions from F020 (39.5 g) (see Table III) to Brine Shrimp

Pool No.	Fractions pooled	Weight (g)	Brine shrimp LC_{50} (mcg/ml)
AT 5	1-5	2.71	>200
AT 10	6-10	1.52	>200
AT 15	11-15	0.64	>200
AT 22	16-22	0.35	>200
AT 25	23-25	0.33	>200
AT 32	26-32	2.70	>200
AT 36	33-36	2.04	0.15
AT 42	37-42	2.31	0.07
AT 49	43-49	2.33	0.13
AT 55	50-55	5.12	0.85
AT 58	56-58	0.66	1.07
AT 67	59-67	1.59	0.33
AT 71	68-71	0.45	0.49
AT 74	72-74	2.46	0.52
AT 80	75-80	2.86	0.37
AT 88	81-88	0.47	0.52
AT 93	89-93	0.38	0.46
AT 98	94-98	2.59	>200
AT 121	99-121	6.12	>200

Table V. Toxicity of Column Chromatography Fractions from AT 49 (1.924 g) (see Table IV) to Brine Shrimp

Solvent	Weight (mg)	Fraction No.	Brine Shrimp LC_{50} (mcg/ml)
50% MeOH (100 ml)	36.6	AT 49-1	0.30
60% MeOH (100 ml)	30.8	AT 49-2	6.28
70% MeOH (100 ml)	28.9	AT 49-3	4.92
80% MeOH (100 ml)	49.2	AT 49-4	0.37
90% MeOH (100 ml)	832.3	AT 49-5	0.36
MeOH (100 ml)	433.5	AT 49-6	0.36
MeOH (100 ml)	363.8	AT 49-7	1.00

Table VI. Chromatotron Separation of AT 49-5 (see Table V)

Pool No.	Fractions	Weight (mg)
AT 49-5-1	1-6	25.9
AT 49-5-2	7-13	342.1
AT 49-5-3	14-16	136.7
AT 49-5-4	17-20	58.3
AT 49-5-5	21-25	20.4

indicating homogeneity. The collected compound, labeled AT-II (81.35 mg), was then subjected to structural analyses, including: high resolution FAB ms, ir, uv, ^{1}H nmr ($CDCl_3$) with selective ^{1}H-^{1}H decouplings, and ^{13}C nmr . This compound was named "asimicin" and was assigned the structure indicated. The structural elucidation of asimicin, without stereochemistry, has been published (23); however, it has eight chiral centers. We have recently defined the stereochemistry at $C_{15,16,19,20,23,24}$ as th/t/th/t/th by comparing ^{1}H nmr values of the triacetate with a series of reference acetylated bis-tetrahydrofurans (24). The configuration at C_4 and C_{36} is suggested to be S and R, respectively, by cd and comparison with rolliniastatin (Hui, Y.-H.; Rupprecht, J.K.; Anderson, J.E.; Liu, Y.-M.; Smith, D.L.; C.-j. Chang; McLaughlin, J.L. J. Nat. Prod., submitted for publication).

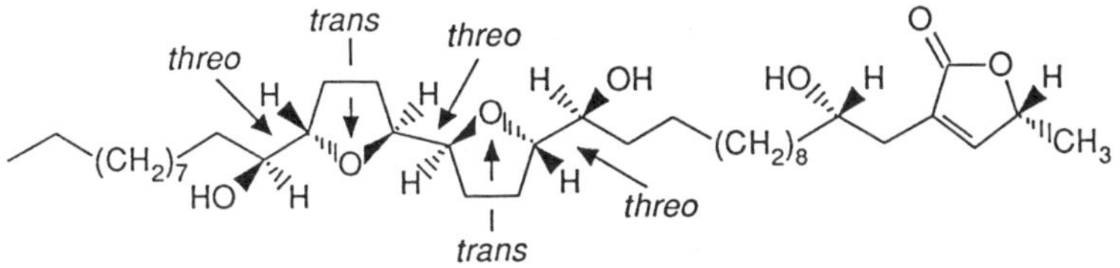

BIOASSAYS OF PAW PAW FRACTIONS AND ASIMICIN

Fractions F017, F020, AT 36, and AT 49 from the fractionation of *Asiminia triloba* bark extract were bioassayed at various levels of test material with several common agronomic pests.

The southern armyworm (*Spodoptera eridania*) bioassay was conducted by first applying an aqueous solution of 5000 ppm test material to leaves of a squash plant and allowing the leaves to dry. The leaves were then removed from the plant and placed in petri dishes with the armyworm larvae. The percent mortality was computed from the number of dead larvae after 3 days.

The effectiveness of the test fractions against the two spotted spider mite (*Tetranychus urticae*), an arthropod, and the melon aphid (*Aphis gossypii*) was determined by spraying infested leaves of squash plants with aqueous solutions containing 4000 ppm test material and observing the percent mortality after 24 hrs.

The assay with yellow fever mosquito larvae, *Aedes aegypti*, involved suspending the larvae in an aqueous solution of the test material, at concentrations of 1000 ppm and less, and determining the percent mortality after 24 hrs.

For assaying with the southern corn rootworm, *Diabrotica undecimpunctata*, soil samples treated with 300 ppm of test material were placed in conical paper cups having a small opening at the lower apical end. Larvae emerging from the opening were evaluated for percent mortality after 3 days.

In the corn leafhopper (*Dalbulus maidis*) assay, filter paper in the bottom of a petri dish was wetted with the test solution and allowed to dry. Thereafter, the leafhoppers and a measured amount of food were introduced to the dish, and the percent mortality was determined after 24 hrs.

The comparative results of the screening of Fractions F017, F020, AT 36, and AT 49 (see Tables III and IV) against the aforementioned test organisms are reported in Table VII. Significant activities were detected against melon aphid and mosquito larvae, some activity was apparent against two spotted spider mites and corn leafhoppers, and no activity was observed against southern armyworm and corn rootworm.

Fractions F017, F018, F019, and F020, from the fractionation of *Asimina triloba* bark extract (see Table III) were bioassayed with mosquito larvae, southern corn rootworm, southern armyworm, two spotted spider mite, and melon aphid. Generally, brine shrimp lethality paralleled the insecticidal activities. In addition, these fractions were assayed with blowfly larvae (*Colliphora vicina*) and the nematode (*Caenorhabditis elegans*).

The blowfly larvae assay was conducted by dipping a gauze dental wick into bovine serum containing 1% (w/v) test material. The larvae were thereafter introduced to the wick and evaluated for percent mortality after 24 hrs.

The nematode (C. elegans) bioassay was conducted by suspending the worms in a 0.1% (w/v) aqueous solution of the test material and determining the percent mortality after 3 days.

The results of this bioassay series are reported in Table VIII.

The activities of Fractions F020, AT 49, and purified asimicin were compared to the commercial pesticides pyrethrum (57%) and rotenone (97%) in assays with the Mexican bean beetle (*Epilachna varivestis*), the melon aphid, mosquito larvae, the nematode C. elegans, and blowfly larvae.

The Mexican bean beetle assay was conducted by spraying the material onto the bean leaves, allowing the leaves to dry, and then introducing the third-instar beetles to the leaves in an enclosed chamber. After 72 hrs., the percent mortality was determined. The assays on the remaining systems were conducted as previously described. The results are reported in Table IX.

Asimicin isolated from the seeds of *Asimina triloba* was assayed by two separate methods against the striped cucumber beetle (*Acalymma vittatum*) as described below.

<u>Two-Choice Leaf Disc Bioassays</u>. Appropriate quantities of the test material were suspended in acetone, diluted with water containing 0.01% of Tween 20 to a sample concentration of 0.5% (w/v), and then the suspension was emulsified with a Brinkman Polytron homogenizer. The 0.1% solutions were derived by dilution of the 0.5% mixture. Leaf discs (2.0 cm diameter) cut from BURPEE HYBRID

cantaloupe leaves were dipped in either the sample homogenate or a corresponding homogenate containing no sample (control discs), air dried, and then two of each type of disc were arranged alternately around 93 mm diameter x 73 mm deep polyethylene dishes. Five newly emerged female striped cucumber beetles, after being starved for 24 hrs, were introduced into each chamber, and the chambers were covered with muslin; these covers were kept moist during the first 6 hrs of the bioassay by misting them periodically with water. Tests were conducted in two replicates under ambient greenhouse conditions. Observations were made at 3, 6, and 22 hrs to estimate visually the amount of leaf tissue consumed and to check for deaths.

Data are presented (Table X) as a consumption index, which is defined as percentage of treated discs consumed x 100/(percentage of control discs consumed+percentage of treated discs consumed). A value of 50 indicates treated and untreated discs have been consumed in equal amounts; an extract that gives an index of 20 or less is considered highly deterrent in these bioassays.

No-Choice Leaf Disc Bioassay. For the no-choice feeding study, homogenate samples preparation (0.5% w/v solutions only), leaf disc (1.5 cm diameter) treatment and drying, and insect preparation were as described for the two-choice bioassay. Single treated discs were placed in individual 2-dram glass vials, and one beetle was introduced into each vial. Water was provided by a soaked 0.5 cm length of dental wick, and the vials were stoppered with cotton plugs. The bioassays were conducted in 10 replicates under ambient greenhouse conditions; leaf consumption and mortality data were taken daily for 3 days. Data are presented as percent of leaf discs consumed.

The results of the two-choice and no-choice leaf disc bioassays are reported in Tables X and XI.

COMMERCIAL POTENTIAL

The chemical synthesis of asimicin would be difficult (there are 256 isomers); thus, a crude extract of the plant material, likely the bark of Annonaceous trees, might furnish an economical source of such acetogenins as pesticides. F020 from the paw paw bark is the most potent of the crude extracts (Tables III, VII, VIII, and IX); a fraction similar to F020 could be readily prepared from the dried, bioactive, bark of a number of Annonaceous trees. The extracts could be standardized by a bioassay, such as brine shrimp lethality, to provide a consistent product. Chemically F020 from paw paw contains asimicin plus at least five additional acetogenins three of which have been isolated and remain to be chemically characterized. We have bioassayed bark extracts of numerous Annonaceae species and find similar activity in many members of this family. TLC suggests that acetogenins are commonly present in the active species.

In a preliminary field test, F020 was suspended in 0.5 or 1% (5,000 or 10,000 ppm) aqueous solutions using 2 or 3% Tween 80 as the suspending agent. The solutions were sprayed onto rows of green bush beans (blue lake variety) which were experiencing a natural infestation with bean leaf beetles (*Cerotoma trifurcata*).

Table VII. Assay of Paw paw (*Asimina triloba*) Fractions for Toxicity to Agronomic Pests

Fraction	Concentration (ppm)	% Mortality Southern armyworm	Two spotted spider mite	Melon aphid	Mosquito larvae
F017	200	10	0	0	---
	20	--[a]	---	---	50
	12	---	---	---	---
F020	200	0	0	30	---
	20	---	---	---	100
	12	---	---	---	---
AT 36	200	0	10	90	---
	100	---	---	---	---
	20	---	---	---	100
	12	---	---	---	---
AT 49	200	0	10	90	---
	100	---	---	---	---
	20	---	---	---	100
	12	---	---	---	---

Fraction	Concentration (ppm)	% Mortality Southern corn rootworm	Corn leafhopper
F017	200	--[a]	---
	20	---	---
	12	0	---
F020	200	---	---
	20	---	---
	12	0	---
AT 36	200	---	---
	100	---	10
	20	---	---
	12	0	---
AT 49	200	---	---
	100	---	0
	20	---	---
	12	0	---

[a]"---" indicates not tested.

Table VIII. Toxicity Assay of Paw paw (*Asimina triloba*) Fractions (Table III) with Representative Agronomic Pests

	% Mortality						
	Mosquito larvae				Blowfly		Southern corn
Fraction	1000 ppm	100 ppm	10 ppm	1 ppm	larvae 1%	*C. elegans* 1%	rootworm 300 ppm
F017	100	90	0	0	0	100	0
F018	50	0	0	0	0	0	0
F019	100	100	70	0	100	100	0
F020	100	100	100	80	100	100	0

	% Mortality		
Fraction	Southern armyworm 5000 ppm	Two spotted spider mite 5000 ppm	Melon aphid 5000 ppm
F017	0	30	0
F018	0	0	0
F019	0	40	20
F020	0	60	50

Table IX. Comparative Activities of Paw paw (*Asimina triloba*) Fractions, Asimicin, and Standard Insecticides

Treatment material	Rate ppm	% Mortality[a]: Mexican bean beetle 72 hr	Melon aphid 24 hr	Mosquito larvae 24 hr	Nematode 72 hr	Blowfly larvae 24 hr
F020	5000	100	80	---	---	---
(Table III)	1000	100	0	100	100	100
	500	100	0	---	---	---
	100	60	0	100	100	0
	10	---	---	80	100	0
	1	---	---	10	0	---
AT 49	5000	100	90	---	---	
(Table IV)	1000	100	0	100	100	100
	500	100	0	---	---	---
	100	100	0	100	100	0
	10	---	---	100	100	-
	1	---	---	50	100	---
Asimicin	1000	---	---	---	---	100
(purified)	500	100	100	---	---	---
	100	100	20	100	100	0
	50	100	0	---	---	---
	10	70	0	100	100	0
	1	0	0	100	100	0
	0.1	---	---	0	100	---
Pyrethrum	500	100	100	---	---	---
(57% pure)	100	100	100	100	0	100
	50	100	100	---	---	---
	10	0	20	100	0	0
Rotenone	1000	---	---	---	---	100
(97% pure)	500	---	0	---	---	---
	100	---	---	100	0	---
	10	---	---	50	---	---
	1	---	---	0	---	---

[a]"---" indicates not tested.

After three applications, within a period of 10 days, there was no visible damage to the new leaves of the test row and it was several inches higher than the control row. The activity seems to remain stable in sunlight. Mechanism of action and toxicology studies remain to be completed.

A U.S. patent has been granted to us covering the use of these Annonaceous acetogenins in pest control (25). A divisional patent on the composition of matter of asimicin is pending. We hope to

Table X. Striped Cucumber Beetle Two-Choice Assay for Asimicin

Asimicin concentration, %	Consumption index[a]		
	3 hr	6 hr	22 hr
0.1	25	20	14
0.5	0	0	0

[a]Value of 50 means that equal amounts of treated and control discs were consumed. Value of 0 means that none of the treated disc was consumed.

Table XI. Striped Cucumber Beetle No-Choice Assay for Asimicin

Treatment	Leaf consumed, %			Mortality, %		
	1 day	2 days	3 days	1 day	2 days	3 days
Asimicin (0.5%)	0	0	0	40	50	50
Control	31	59	74	0	0	0

license this discovery as an "all natural" garden pesticide. While our patent was being processed, another U.S. patent (26) was issued to the Bayer Co. in Germany; their patent protects the insecticidal use of an unknown substance, called "annonin", isolated from *Annona squamosa* (Annonaceae).

CONCLUSIONS

The screening of higher plants for novel pesticides detects interesting active leads. The activity-directed fractionation of seeds of *Thevetia thevethioides* (Apocynaceae), using European corn borers, led us to two known cardiac glycosides, neriifolin and 2'-acetylneriifolin. The activity-directed fractionation of seeds and bark of *Asimina triloba* (Annonaceae) (paw paw) led us to a mixture of potent novel acetogenins of which asimicin has been characterized. These acetogenins are common to a large number of Annonaceae. Commercial development of the Annonaceous extracts as a garden pesticide seems feasible provided that the materials prove safe. In addition to being pesticidal, we have determined that asimicin has additional biological effects which may be worthy of further study. The fractionation of target plant species in other families should provide more of such new compounds for practical exploitation.

LITERATURE CITED

1. Jacobson, M. Insecticides of the Future; Marcel Dekker: New York, 1975.
2. Mandava, N.B., ed. Handbook of Natural Pesticides: Methods; CRC Press: Boca Raton, FL, 1985; Vols. 1 and 2.
3. Leahey, J.P. The Pyrethroid Insecticides; Taylor and Francis, Inc.: Philadelphia, 1985.
4. Jacobson, M.; Crosby, D.G. Naturally Occurring Insecticides; Marcel Dekker: New York, 1971.
5. Balandrin, M.F.; Klocke, J.A.; Wurtele, E.S.; Bollinger, W.H. Science 1985, 228, 1154.
6. Schroeder, D.R.; Nakanishi, K. J. Nat. Prod. 1987, 50, 241.
7. Jacobson, M. Insecticides from Plants, A Review of the Literature, 1941-1953; USDA Agric. Handb. 154, 1958; p 299.
8. Jacobson, M. Insecticides from Plants, A Review of the Literature, 1954-1971; USDA Agric. Handb. 461, 1975; p 138.
9. Guthrie, W.D.; Raun, E.S.; Dicke, F.F.; Pesko, G.R.; Carter, S.W. Iowa State J. Sci. 1965, 40, 65.
10. Freedman, B.; Nowack, L.J.; Kwolek, W.F.; Berry, E.C.; Guthrie, W.D. J. Econ. Entomol. 1979, 72, 541.
11. McLaughlin, J.L.; Freedman, B.; Powell, R.G.; Smith, C.R., Jr. J. Econ. Entomol. 1980, 73, 398.
12. Freedman, B.; McLaughlin, J.L.; Powell, R.G., Smith, C.R., Jr.; Reed, D.K.; Ledd, T.L., Jr. Control of Insect Pests with Neriifolin and 2'-Acetylnerrifolin; Serial No. 061289,948 filed for U.S. Patent Aug. 4, 1981; abandoned June 29, 1982.
13. Jaspers, E., ed. Fundamental and Applied Research on The Brine Shrimp Artemia salina (L.) in Belgium; European Mariculture Society, Special Publication No. 2: Bredene, Belgium, 1977.
14. Meyer, B.N.; Ferrigni, N.R.; Putnam, J.E.; Jacobsen, L.B.; Nichols, D.E.; McLaughlin, J.L. Planta Medica 1982, 45, 31.
15. Anderson, J.E.; Chang, C.-j.; McLaughlin, J.L. J. Nat. Prod. 1988, 51, 307.
16. Alkofahi, A.; Rupprecht, J.K.; Smith, D.L.; Chang, C.-j.; McLaughlin, J.L. Experientia 1988, 44, 83.
17. Kubo, I.; Lee, Y .-W.; Pettei, M.; Dilkiewicz, F.; Nakanishi, K. J.C.S. Chem. Comm. 1976, 1013.
18. Nakanishi, K.; Kubo, I. Israel J. Chem. 1977, 16, 28.

19. Kubo, I.; Miura, I.; Pettei, M.; Lee, Y.-W.; Dilkiewicz, F.; Nakanishi, K. Tetrahedron Lett. 1977, 52, 4553.

20. Kubo, I.; Matsumoto, T.; Kakooko, A.; Mubiru, N. Chem. Lett. 1983, 979.

21. Nakata, T.; Akita, H.; Naito, T.; Oishi, T. J. Am. Chem. Soc. 1979, 101, 4400.

22. Okawara, H.; Nakai, H.; Ohno, M. Tetrahedron Lett. 1982, 1087.

23. Rupprecht, J.K.; Chang, C.-j.; Cassady, J.M.; McLaughlin, J.L.; Mikolajczak, K.L.; Weisleder, D. Heterocycles 1986, 24, 1197.

24. Hoye, T.R.; Sukadolnik, J.C. J. Am. Chem. Soc. 1987, 109, 4402.

25. Mikolajczak, K.L.; McLaughlin, J.L.; Rupprecht, J.K. U.S. Patent 4 721 729, issued Jan. 26, 1988.

26. Moeschler, H.G.; Pfluger, W.; Wendisch, D. U.S. Patent 4 689 232, issued Aug. 25, 1987.

RECEIVED November 28, 1988

Chapter 4

Toxicity and Fate of Acetylchromenes in Pest Insects

Murray B. Isman

Department of Plant Science, University of British Columbia, Vancouver, British Columbia V6T 2A2, Canada

Acetylchromenes (benzopyrans), major natural constituents of the desert sunflower *Encelia*, are toxic to a wide range of insects via residue contact. Bioassayed on glass surfaces, the predominant compound, encecalin, has LC_{50}s of 1.4 and 11.8 $\mu g/cm^2$ against the variegated cutworm (Noctuidae), and the migratory grasshopper (Acrididae), respectively. Structure-activity relations indicate that chromene derivatives possessing free hydroxyl groups or saturated heterocycles are significantly less toxic than encecalin. Acetylchromenes applied topically to adult grasshoppers are rapidly absorbed from the cuticle, and both metabolites and parent chromenes appear in the excreta within two hours of application. However, only 20-40% of the applied dose is excreted, and the remainder cannot be recovered by extraction of the insect. Simple metabolites result from oxidation of the geminal methyl groups or reduction of the ketone. Subsequent bioassays of these metabolites, either by residue contact or topical application, indicate that they are significantly less toxic than their parent acetylchromenes. Benzofurans, closely related constituents of *Encelia* and other desert plants, are non-toxic to insects. However, benzofurans antagonize the toxicity of encecalin in three species of insects tested to date. The benzofuran euparin did not interfere with the antihormonal action of precocene III in the milkweed bug. Encecalin exhibits substantial antifeedant activity against several species of noctuid caterpillars.

It is well accepted that natural products from plants may constitute new sources of pest control materials or prototypes for such materials, and to that end there is a current worldwide effort aimed at screening hitherto unexamined plant species for bioactivity towards pest insects (*1-2*). In particular, tropical plants have proven to be a rich source of bioactive chemicals. Plants from arid and semi-arid regions have also been singled out as having a high potential for the discovery of natural insecticides and

0097–6156/89/0387–0044$06.00/0

antifeedants (*3*), but to date, little emphasis has been placed on investigations of the phytochemistry and bioactivity of desert plants (*4*).

One group of plants native to the New World deserts which possess insecticidal constituents are desert sunflowers of the genus *Encelia* (Asteraceae). All of the twenty or so taxa in this genus elaborate as major constituents benzopyrans (chromenes) and benzofurans (*5*). The present paper will review the bioactivity of these natural products in both pest and non-pest insects.

Natural Products from *Encelia*

Most species of *Encelia* elaborate three major acetylchromenes (6-acetyl-2,2-dimethyl-benzopyrans) (Figure 1: compounds 1,4 and 5), two benzofurans (compounds 12,13) and stereoisomers of an unusual dimer (compound 11) which represents a condensation of encecalin (compound 1) and euparin (compound 13), the principal chromene and benzofuran, respectively (*5*). These constituents occur exclusively in the resin ducts of the plant (*6*), and there is wide intra- and inter-specific variability in their concentrations (*5*, and P. Proksch, unpublished data). Encecalin concentrations can reach 150 μmol/g dry weight (approximately 3.5% dwt) in *E. farinosa* (*5*) and can constitute over 90% of the total chromenes/ benzofurans, for example in *E. palmeri* (P. Proksch, unpublished data). As these compounds occur exclusively in resin ducts, the greatest concentrations within plants are found in the stems and foliage compared to the floral parts and roots (*7*). *Encelia* species tend not to be attacked by insect herbivores in their natural habitat, but notable exceptions are the specialist leaf beetle *Trirhabda geminata* (Chrysomelidae) (*8*), and the generalist grasshopper *Cibolacris parviceps* (Acrididae)(R. Chapman, pers. comm.).

Bioactivity of Chromenes and Benzofurans

Attention to chromenes amongst entomologists was first generated over a decade ago by the discovery that the precocenes, 7-methoxy-chromenes (Figure 1, compound 9) isolated from the ornamental plant *Ageratum houstonianum*, induce precocious molting in the milkweed bug, *Oncopeltus fasciatus* (Hemiptera), by a specific cytotoxic action on the corpora allata, the glands which synthesize juvenile hormone (*9, 10*). However, the precocenes are not representative of the thirty or so more widely distributed naturally-occurring chromenes, which are characterized by the acetyl substitution at position 7 (e.g. encecalin). Because the structure of encecalin differs from that of precocene II only at the 7 position, it was tested several years ago for anti-juvenile hormone activity, but found to be lacking any activity in this regard (*11*, and W.S. Bowers, pers. comm.).

Antifeedant Effects. Encecalin, added to artificial media at concentrations of 15 μmol/g fwt, completely deterred neonate bollworms (*Heliothis zea*, Noctuidae) to starvation (*12*). A similar effect was seen with neonate loopers (*Plusia gamma*) at 3 μmol/g fwt; at 18 μmol/g fwt the feeding of fourth instar loopers was reduced by 65% relative to controls (*13*). A similar study using the variegated cutworm, *Peridroma saucia*, another noctuid pest, produced 85% mortality of neonates when placed on media spiked with encecalin at 3 μmol/g fwt (*13*). This last result provided the impetus to determine if mortality resulted from starvation or a more direct insecticidal action.

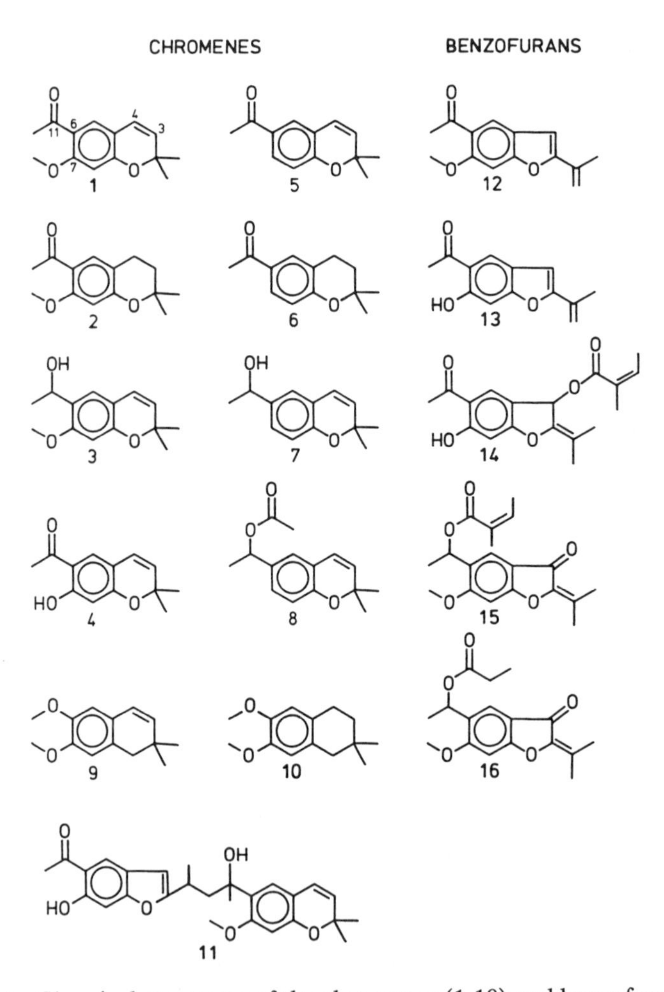

Figure 1. Chemical structures of the chromenes (1-10) and benzofurans (12-16) bioassayed against the variegated cutworm and the migratory grasshopper. Reproduced with permission from Ref. 15. Copyright 1987, Dr W. Junk Publishers.

Crude extracts of *Encelia* species were bioassayed for antifeedant properties against fifth instar *P. saucia* by adding methanolic extracts from four species to cabbage leaf discs at natural concentrations in a choice test. Disc area consumed was estimated by eye using a semi-quantitative scale wherein a score of 1 = 0-25% consumption, 2 = 26-50% and so on. There were ten larvae per replicate and the experiment was replicated four times. The results (Table I) indicate a strong correlation ($r^2 = 0.87$) between inhibition of feeding and encecalin content of the crude plant extracts.

Table I. Feeding Response of 5th Instar Variegated Cutworms to Methanolic Extracts of *Encelia* Species Coated Onto Cabbage Leaf Discs

Species	Encecalin Content (mg/g dwt)	Mean Feeding Score[a]	
		2 hr	6 hr
Solvent CONTROL	0.0	2.1 a	2.4 a
E. actoni	0.1	1.7 ab	2.2 ab
E. farinosa (Sonora)	0.7	1.6 ab	2.6 a
E. farinosa (Mohave)	5.3	1.2 b	1.6 bc
E. palmeri	8.5	1.1 b	1.4 c

Means within a column followed by the same letter are not significantly different (p=.05, Fisher's LSD test); n=10, with four replicates.

[a]Visually estimated as follows: 1=0-25% consumed; 2=26-50%; 3=51-75%; 4=76-100%

Contact Toxicity. To assess the direct contact insecticidal action of chromenes and benzofurans, pure compounds were dissolved in a volatile carrier solvent and coated onto the inner surfaces of 20 mL glass vials. After the carrier had evaporated, neonate test insects were placed in the vials with food and survival assessed at 24-48 hr (*13-14*). This method demonstrated unambiguously that encecalin was lethal to neonate variegated cutworms with an LC_{50} of 0.22 μmol/vial which is equivalent to 1.15 μg/cm^2 (*15*). A dose of 1.0 μmol placed in the cap rather than coated onto the walls of the vial did not give rise to significant mortality, strongly indicating that toxicity resulted from direct contact rather than volatile action (*13*). Using neonate migratory grasshoppers (*Melanoplus sanguinipes*, Acrididae), the same bioassay yielded an LC_{50} for encecalin of 1.5 μmol/vial (*14*). However, neonate hoppers weigh approximately fifty times more than neonate cutworms, suggesting that the former insect is much more susceptible to this compound. In contrast, concentrations of encecalin exceeding 20 μg/cm^2 were required to kill neonate milkweed bugs (11).

Structure-Activity Relations. Using the above-mentioned bioassay, a series of seven naturally occurring and four synthetic chromenes were assessed for toxicity to neonate cutworms, as well as five benzofurans (*15*). None of the benzofurans proved to be significantly toxic at the initial screening concentration of 1.0 μmol/vial (approximately 5 μg/cm^2). Of the chromenes tested, the most potent was the allatocidin, precocene II. Analogues possessing free hydroxyl groups (compounds 3,4 and 7) were significantly less active, as were dihydro analogues where the heterocycle was saturated (compounds 2,6 and 10). It is noteworthy that the unsaturation in the heterocycle is essential for the anti-hormonal action of the precocenes, and appears to be important in the insecticidal action of the precocenes and acetylchromenes. In addition to the lethal action of the acetylchromenes, surviving larvae in a residue contact bioassay utilizing 2-day-old larvae grew significantly more poorly over 8 days following a brief (24 hr) exposure to residues of encecalin or its demethoxy analogue, indicating sublethal toxicity (*15*).

Of particular interest with respect to structure-activity relations is the observation that the demethyl analogue of encecalin (Figure 1: compound 4) is significantly less active (via residue contact) than encecalin in each of four species of insects representing four insect orders (Table II). Although demethylencecalin possesses a free hydroxyl group, it is less polar than encecalin owing to internal hydrogen bonding of the hydroxyl with the keto oxygen (*16*). As the consistent difference in toxicity of the two structurally similar compounds might reflect differential pharmacokinetics of the compounds, this hypothesis became the subject of an investigation which is reviewed in the next section.

Table II. Toxicity of Encecalin and Demethylencecalin to Insects: LC_{50} via Residue Contact

Insect	Encecalin(1)	Demethylencecalin(2)	Reference
Oncopeltus fasciatus (Hemiptera: Lygaeidae)	20 μg/cm^2	>80 μg/cm^2	*11*
Culex pipiens (Diptera: Culicidae)	3.0 ppm	6.4 ppm	*16*
Peridroma saucia (Lepidoptera: Noctuidae)	1.3 μg/cm^2	>4.5 μg/cm^2	*15*
Melanoplus sanguinipes (Orthoptera: Acrididae)	12 μg/cm^2	>22 μg/cm^2	*17*

Numbers in parentheses refer to structures in Figure 2.

The residue contact LC_{50} for demethylencecalin has been impractical to establish for neonate grasshoppers because large quantities of this scarce chemical are required to coat the glass vials. To precisely assess the toxicities of encecalin and its demethyl and demethoxy analogues to the migratory grasshopper, these compounds were topically applied to neonate hoppers using 0.5 μL of acetone as the carrier. Mortality was assessed 48 hr post-treatment; hoppers were fed fresh wheat blades and held at 27°C. Direct topical application of encecalin and demethoxyencecalin yielded highly linear dose-responses (Figure 2). Although results are variable from replicate to replicate, relationships were linear within every replicate, and pooling of the replicates indicates that the two compounds are equitoxic with LD_{50}s of approximately 66 nmol/insect (15,13 μg respectively). Surprisingly, via topical application demethylencecalin was found to be more toxic than the other two acetylchromenes (Figure 2). Several replicates indicated that the dose-response for demethylencecalin is non-linear; interpolation of the curvilinear response suggests an LD_{50} of approximately 38 nmol/insect.

It is presently unclear as to why this last result runs contrary to the result for the residue-contact bioassay utilizing the same test species and other species as indicated in Table II, but it is noteworthy that demethylencecalin is much less volatile than the other two acetylchromenes. At room temperature demethylencecalin is a solid whereas encecalin and its demethoxy analogue are oils. Thus a residue-contact bioassay would favor uptake of residues of the latter two compounds. However, topical application of encecalin and its demethyl analogue to third instar milkweed bugs indicated that encecalin is significantly more toxic (*16*), a result confirmed in my laboratory (unpublished data), and in agreement with the differential toxicity summarized in Table II.

Metabolism and Fate of Acetylchromenes

The pharmacokinetic and metabolic fates of the major *Encelia* acetylchromenes were investigated in adult migratory grasshoppers following topical administration. Both encecalin and demethylencecalin are rapidly absorbed from the cuticle; in both cases approximately 55% of the applied dose was absorbed in the first 2 hours posttreatment, with over 95% absorbed after 24 hours (Figure 3; shows results for encecalin only) (*17*). Similarly, when pure encecalin was applied topically to fifth instar variegated cutworms, 75% of the applied dose was absorbed at 2 hours, and no residual encecalin could be recovered from the cuticle after 24 hours (unpublished data).

Encecalin, demethylencecalin, and their respective major metabolites (quantified by high performance liquid chromatography) appear in the frass within 2 hours of treatment. Maximal excretion of both parent chromenes and metabolites from 4-8 hours posttreatment, and there is minimal excretion beyond 12 hours (Figure 3; shows results for encecalin only) (*17*). The timecourse for excretion in the cutworm is similar except that maximal excretion occurs between 2 and 4 hours posttreatment.

In the grasshopper, quantification of excreted chromenes and metabolites from several groups of treated insects indicates that only 20-40% of the applied dose is excreted (Table III), almost all of which is excreted with 24 hours of administration. In the case of encecalin and demethoxyencecalin, more than 80% of that which is excreted is in the form of metabolites, whereas for demethylencecalin two-thirds of the excreted dose is as the parent compound. In the cutworm, only 15% of the applied dose of encecalin was excreted in the frass. This occurred exclusively as the parent compound; there was no evidence of metabolites as seen in the grasshopper excreta.

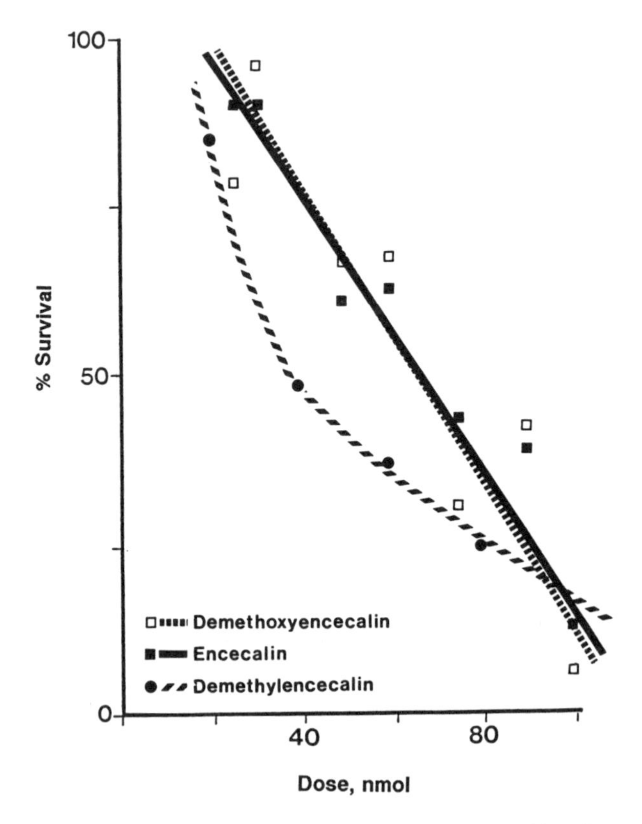

Figure 2. Dose-response relationships for the major *Encelia* acetylchromenes to *Melanoplus* neonates via topical application. Encecalin (■); demethoxyencecalin (□); demethylencecalin (●). For encecalin and demethoxyencecalin, each point represents 8 groups with 5 insects per group (n=40); for demethylencecalin, each point represents 24-40 groups (n=120-200). For the first two named compounds, lines fitted by linear regression. For demethylencecalin, line fitted by eye.

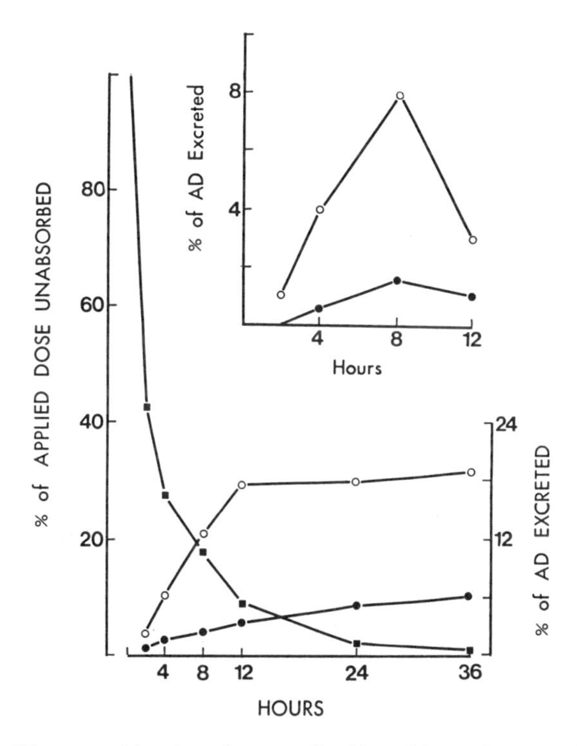

Figure 3. Pharmacokinetics of encecalin (●) and its major metabolite (○) in the adult migratory grasshopper. Encecalin recovered (■) in external rinse; not differences in scale for unabsorbed and excreted chromenes. Data for excreted chromenes are cumulative. Inset: results from a continuously sampled cohort. Reproduced with permission from Ref. 17. Copyright 1987, Alan R. Liss, Inc.

Table III. Excretion of Acetylchromenes by the Migratory Grasshopper 24 Hours After Topical Application

Treatment[b]	n[c]	% of AD[a]	% of ED[a] as parent
		mean ± SD	mean ± SD
Demethoxyencecalin(5)	5	19.2 ± 8.5	13.5 ± 2.6
Encecalin(1)	8	30.3 ± 7.8	18.3 ± 11.0
Demethylencecalin(4)	8	40.4 ± 8.4	66.1 ± 5.5

[a] AD = applied dose (60 μg/insect); ED = excreted dose (parent compound metabolites).
[b] Numbers in parentheses refer to structures in Figure 2.
[c] n = no. of treated groups with four insects per group.

The migratory grasshopper excretes simple metabolites of the acetylchromenes; there is no evidence for conjugated metabolites as has been reported for precocene II in this insect (*18*). Following treatment with demethylencecalin a single metabolite was found, the result of hydroxylation of one of the gem-methyl groups (Figure 4: compound 4a). Demethoxyencecalin gave rise to three metabolites, the most prominent being the 2-hydroxymethyl derivative (compound 5a) of the parent. This compound also occurs as a natural product in the plant *Ageratina altissima*, and is produced as a metabolite of demethoxyencecalin when the acetylchromene is applied to liquid cell suspension cultures of this plant (P. Proksch, unpublished data). Encecalin gave rise to seven metabolites (Figure 5), but as in the case for the other two acetylchromenes the predominant metabolite is the 2-hydroxymethyl derivative (compound 1a). Two of the metabolites of encecalin are chromans (the heterocycle is saturated); chromans have not previously been reported as natural metabolites of chromenes (e.g. precocenes) from insects.

A different pathway of metabolism in both encecalin and demethoxyencecalin arises from reduction of the ketone to an alcohol (compounds 1f and 5b). Encecalin is also demethylated by the insect. These metabolites can be further hydroxylated on one of the gem-methyl groups (compounds 1e, 1g, 5c). As the predominant metabolites of all three of the major acetylchromenes are the 2-hydroxymethyl derivatives, and the secondary metabolites can serve as substrates for this reaction, it appears that aliphatic hydroxylation of the gem-methyl groups is the major mode of metabolism of these compounds in this insect.

It is interesting to note that diols, major metabolites of precocenes reported from several insect species (*19*), were absent in our preparations. This was true not only for the acetylchromenes, but also in the case of precocene II when applied to grasshoppers.

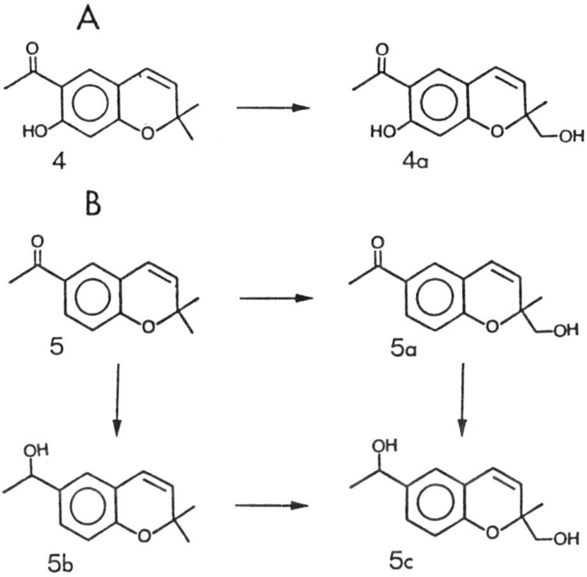

Figure 4. Metabolic fate of demethylencecalin (A) and demethoxyencecalin (B) in the migratory grasshopper. Reproduced with permission from Ref. 17. Copyright 1987, Alan R. Liss, Inc.

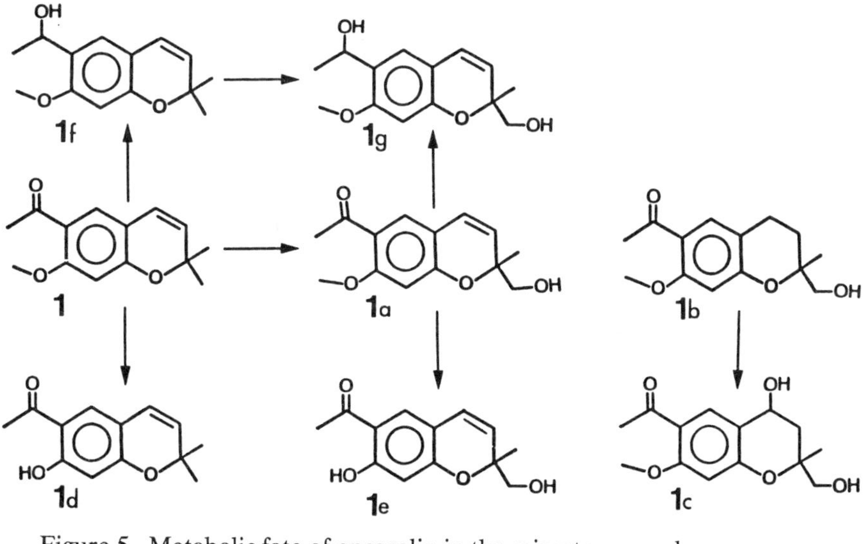

Figure 5. Metabolic fate of encecalin in the migratory grasshopper. Reproduced with permission from Ref. 17. Copyright 1987, Alan R. Liss, Inc.

What is the fate of the remaining 60-80% of the applied acetylchromene dose which is not excreted? No additional chromenes or metabolites could be recovered from the grasshoppers by extraction of carcass homogenates either before or after hydrolysis. This suggests that the parent chromenes alkylate macromolecules in the insect, becoming irreversibly bound. Provided that radiolabelled acetylchromenes can be prepared (in progress), it should be possible to establish the ultimate fate of the acetylchromenes in the grasshopper and other insects. Such findings should shed some light on the mechanism-of-action of these compounds, particularly as the mechanism may be unrelated to the allatocidal action of the precocenes.

Toxicity of Metabolites. Hydroxylation of acetylchromenes as a major mode of metabolism by the grasshopper is advantageous in two ways. Acetylchromenes possessing free hydroxyl groups (with the exception of demethylencecalin) are more polar than their parent chromenes, and are presumably more amenable to excretion by the Malpighian tubules of the insect. In addition to enhancing excretion, hydroxylation also appears to constitute a detoxicative process. Purified metabolites of encecalin and demethoxyencecalin are significantly less toxic to neonate grasshoppers via topical application as seen in Table IV. At a dose of 100 nmol/insect, the 2-hydroxymethyl derivative of demethoxyencecalin is essentially non-toxic, as is the 11-hydroxy derivative of encecalin. The 11-hydroxy derivative of demethoxyencecalin is only slightly active at this relatively high dose. These data permit one to speculate that the relative susceptibility of an insect to the toxic action of acetylchromenes may be a function of the relative ability of that insect to convert the toxic parent compound to non-toxic, hydroxylated metabolites.

Table IV. Toxicity of Chromene Metabolites Topically Applied to Neonate Migratory Grasshoppers

Compound[a]	Dose (nmol)	n[b]	% Survival (48 hr) mean ± SE
Demethoxyencecalin(DMX;5)	100	16	6.3 ± 3.4
2'-OH-DMX (5a)	100	8	95.0 ± 3.3
11-OH-DMX (5b)	100	8	65.0 ± 9.8
Encecalin (1)	100	16	12.0 ± 3.6
11-OH-encecalin (1f)	100	8	90.0 ± 3.8
CONTROL		50	95.6 ± 1.4

[a] numbers in parentheses refer to structures in Figures 4 and 5
[b] no. of vials with 5 insects per vial

Interactions Between Benzofurans and Chromenes

For many years, studies of insect-plant chemical interactions often consisted of the isolation of a purified plant natural product, which, when incorporated into an artificial medium resulted in a behavioral or physiological response from the test insect. It is now well recognized that many natural products which are biologically-active against insects when isolated from the natural context fail to elicit the expected response *in planta* (e.g. *20*). In contrast, some crude plant extracts have considerable bioactivity, but yield isolated compounds which lack activity, although activity is restored or enhanced when the pure components are subsequently mixed and offered as a blend (21). In the latter case, bioactivity depends on the synergistic action of the natural products. Certain plant natural products have long been known to synergize the toxicity of synthetic insecticides(*22*), and recently synergistic effects of co-occurring allelochemicals have been demonstrated (*23-24*).

As the benzofurans (especially compounds 12 and 13, Figure 1) consistently co-occur with the acetylchromenes in *Encelia* species, but the former compounds have little bioactivity themselves, their potential influence on the toxicity of chromenes was investigated. Neonate migratory grasshoppers were treated topically with 90 nmol of encecalin alone, or an equimolar mixture of encecalin with one of four benzofurans or the encecalin/euparin dimer . Surprisingly, all of the benzofurans antagonized the toxicity of encecalin in this bioassay (Table V). Although euparin (compound 13) was among the least active in this regard, it still enhanced survival by 50% relative to insects treated with encecalin alone. Euparin itself was completely inactive (95% survival). Compound 15, an angelate benzofuran ester from *Enceliopsis argophylla* (a close relative of *Encelia*)(*25*), increased survival four-fold, as did the dimer (from *Encelia*).

Table V. Antagonism of Encecalin Toxicity in Neonate Migratory Grasshoppers by Benzofurans Following Topical Application

Compound(s)[a]	Dose (nmol)	n[b]	% Survival (48 hr) mean ± SE
CONTROL		50	95.6 ± 1.4
Euparin (13)	90	8	95.0 ± 3.3
Encecalin (ENC)(1)	90	40	20.5 ± 3.6
ENC + euparin	90 + 90	40	31.5 ± 3.8
ENC + compound 12	90 + 90	24	56.0 ± 5.2
ENC + compound 14	90 + 90	24	29.2 ± 4.3
ENC + compound 15	90 + 90	24	79.2 ± 4.3
ENC + compound 11	90 + 90	8	75.0 ± 9.1

[a] Numbers for compounds refer to structures in Figure 2
[b] no. of vials with 5 insects per vial

These results were corroborated by a further experiment in which encecalin, euparin, and an equimolar mixture of the two compounds were applied topically to third instar milkweed bugs (Table VI). In this case, euparin enhanced survival three-fold.

Preliminary experiments utilizing the residue-contact bioassay method with neonate variegated cutworms or migratory grasshoppers indicated that euparin was capable of antagonizing the insecticidal action of precocene II. Thus, a final experiment was conducted to determine if euparin could antagonize the allatocidal (anti-hormonal) effect of precocene III, a synthetic analogue of the natural precocenes, in the milkweed bug, a species sensitive to the precocenes. Euparin had no influence on the effect of precocene III (Table VI).

Table VI. Influence of the Benzofuran, Euparin, on the Biological Activity of Encecalin and Precocene III in the Large Milkweed Bug

Compound (Dose per insect)	n	% Survival	% Adultoids
CONTROL	40	100	0
Euparin (50 nmol)	40	95	0
Precocene III (20 nmol)	50	92	100
Encecalin (50 nmol)	40	17	0
Encecalin + Euparin (50 + 50)	40	55	0
Precocene + Euparin (20 + 20)	50	96	100

The antagonism of encecalin toxicity by the benzofurans raises interesting questions concerning the role of these putative insecticides/antifeedants in the natural context. An investigation is underway to assess the relationships between acetylchromene/benzofuran concentrations and cutworm feeding, growth and survival utilizing foliar extracts from a range of *Encelia* species. From the practical point of view, however, the benzofurans may be a most useful tool in the ongoing investigation of the mechanism-of-action of the insecticidal acetylchromenes.

Practical Considerations and Prospects

Do the *Encelia* acetylchromenes have potential as commercial insecticides? The answer would appear to be no. These natural products, while exhibiting interesting insecticidal action, are significantly less efficacious (by two or more orders of magnitude) than other botanical insecticides in use or under development (e.g. azadirachtin). Encecalin is also quite susceptible to degradation in sunlight. However, these are properties which could conceivably be altered by synthesis of analogues with enhanced bioactivity and stability. It is noteworthy, however, that efforts to do just that with the precocenes have met with little success (*26-27*). Development of commercial products based on the precocenes was dealt a considerable blow with the discovery that these compounds are potent hepatotoxins in the rat, via the same mechanism which accounts for the allatocidal effect in insects (*28*).

At present, the toxicity of the acetylchromenes to vertebrates is unknown, but it may well be that the insecticidal action results from a general cytotoxic effect, which could preclude development. On the other hand, knowledge obtained from the investigations described in the present review may find practical application in the development of other pest control materials targeted at the pests used herein.

Acknowledgments

The author wishes to thank Dr. Peter Proksch (Braunschweig, West Germany) for generously providing the pure compounds used in these investigations, for permission to cite unpublished results, and for many stimulating discussions. The bioassays involving topical application of compounds to neonate grasshoppers were skilfully conducted by Nancy Brard, whose technical support is most appreciated. Desiree Jans conducted the study of antifeedant effects of *Encelia* extracts presented in Table I.
The author gratefully acknowledges grants received from NSERC (A2729, E6850, and SE0148), and the DFG (444 KAN-11).

Literature Cited

1. Balandrin, M.F.; Klocke, J.A.; Wurtele, E.S.; Bollinger, W.H. Science 1985, 228, 1154-1160.
2. Plimmer, J.R. In Bioregulators for Pest Control; Hedin, P.A., Ed.; ACS Symposium Series No. 276; American Chemical Society: Washington, DC, 1985; pp 323-335.
3. Hedin, P.A. J.Agric. Food Chem. 1982, 30, 201-215.
4. Rodriguez, E. in Plant Resistance to Insects; Hedin, P.A., Ed.; ACS Symposium Series No. 208; American Chemical Society: Washington, DC, 1983; pp 291-302.
5. Proksch, P; Clark, C. Phytochemistry 1987, 26, 171-174.
6. Proksch, P.; Proksch, M.; Weck, W.; Rodriguez, E. Z. Naturforsch. 1985, 40c, 301-304.
7. Proksch, P.; Rodriguez, E. Biochem. Syst. Ecol. 1984, 12, 179-181.
8. Wisdom, C.S.; Rodriguez, E. Biochem. Syst. Ecol. 1983, 11, 345-352.
9. Bowers, W.S.; Ohta, T.; Cleere, J.S.; Marsella, P.A. Science, 1976, 193, 542-547.
10. Brooks, G.T.; Pratt, G.E.; Jennings, R.C. Nature 1979, 281, 570-572.
11. Proksch, P.; Proksch, M.; Towers, G.H.N.; Rodriguez, E. J. Nat. Prod. 1983, 46, 331-334.
12. Wisdom, C.S.; Smiley, J.T.; Rodriguez, E. J. Econ. Entomol. 1983, 76, 993-998.
13. Isman, M.B.; Proksch, P. Phytochemistry 1985, 24, 1949-1951.
14. Isman, M.B.; Yan, J.-Y.; Proksch, P. Naturwissensch. 1986, 73, 500-501.
15. Isman, M.B.; Proksch, P.; Yan, J.-Y. Entomol. exp. appl. 1987, 43, 87-93.

16. Klocke, J.A.; Balandrin, M.F.; Adams, R.P.; Kingsford, E. J. Chem. Ecol. 1985, 11, 701-712.
17. Isman, M.B.; Proksch, P.; Witte, L. Arch. Insect Biochem. Physiol. 1987, 6, 109-120.
18. Bergot, B.J.; Judy, K.J.; Schooley, D.A.; Tsai, L.W. Pestic. Biochem. Physiol. 1980, 13, 95-104.
19. Soderlund, D.M.; Messeguer, A.; Bowers, W.S. J. Agric. Food Chem. 1980, 28, 724-731.
20. Isman, M.B.; Duffey, S.S. J. Amer. Soc. Hortic. Sci. 1982, 107, 167-170.
21. Adams, C.M.; Bernays, E.A. Entomol. exp. appl. 1978, 23, 101-109.
22. Berenbaum, M.R.; Neal, J.J. In Allelochemicals: Role in Agriculture and Forestry; Waller, G.R., Ed.; ACS Symposium Series No. 330; American Chemical Society: Washington, DC, 1987; pp 416-430.
23. Berenbaum, M.R. Rec. Adv. Phytochem. 1985, 19, 139-169.
24. Berenbaum, M.R.; Neal, J.J. J. Chem. Ecol. 1985, 12, 1349-1358.
25. Budzikiewicz, H.; Laufenberg, G.; Clark, C.; Proksch, P. Phytochemistry 1984, 23, 2625-2627.
26. Bowers, W.S. In Insecticide Mode of Action; Coats, J.R., Ed.; Academic Press: New York, NY, 1982; pp 403-427.
27. Camps, F. In Bioregulators for Pest Control; Hedin, P.A., Ed.; ACS Symposium Series No. 276; American Chemical Society: Washington, DC, 1985; pp 237-243.
28. Halpin, R.A.; Vyas, K.P.; El-Naggar, S.F.; Jerina, D.M. Chem.-Biol. Interactions 1984, 48, 297-315.

RECEIVED November 2, 1988

Chapter 5

Nonprotein Amino Acid Feeding Deterrents from *Calliandra*

John T. Romeo[1] and Monique S. J. Simmonds[2]

[1]Department of Biology, University of South Florida, Tampa, FL 33620
[2]Royal Botanic Gardens, Kew, Richmond, Surrey TW9 3AB, England

Several nonprotein imino acids and sulphur amino acids are found in seeds, leaves and sap of Calliandra, a tropical Mimosoid legume. Concentrations may reach 3% dry weight. Individual imino acids exhibit modest insecticidal activity against some lepidopterans, affecting all stages of the life cycle. Varying degrees of mortality are seen in Spodoptera and Heliothis species, and delayed growth and decreased fecundity are also common. S-(β-carboxyethyl)-cysteine and cis-4-hydroxypipecolic acid are deterrent to feeding by these insects. In aphids, combinations of imino and sulphur amino acids, which mimic those found in the plants, are deterrent to feeding, and they also reduce survival and fecundity. Electrophysiological, behavioral, and nutritional studies have not yet clarified the mode of action of these compounds.

The tropical Mimosoid legume genus Calliandra consists of some 200 species confined mostly to the New World. The plants range from woody shrubs to medium size trees, and have not yet been exploited on a large scale. In Indonesia, however, the rapidly growing C. calothyrsus is widely cultivated for firewood. It is recommended as a supplementary plant in villages and rural areas for use as fuel and fodder so that natural forests can be spared destruction. Foliage contains up to 22 percent protein and cattle and goats consume it freely. Bees use its nectar for producing honey (1). While as yet untested, other species are likely to prove useful in the humid tropics, and Calliandra has been designated as one of eight potentially exploitable tropical plant groups needing further scientific investigation (2).

0097–6156/89/0387–0059$06.00/0

Nonprotein Amino Acid Chemistry of Calliandra

The foliage of Calliandra is known to contain an array of rare nonprotein imino acids (amino acids containing a heterocyclic nitrogen ring) which are largely confined to legumes. The compounds are derivatives of pipecolic acid, the higher homologue of proline, and include an acetylamino, four monohydroxy, and four dihydroxy derivatives (Figure 1).

Distribution patterns are species specific and usually the monohydroxy precursors are found together with one or two dihydroxy compounds (3). In Calliandra, pipecolic acid is synthesized from lysine and the imino ring is subsequently hydroxylated to form the mono and dihydroxy derivatives (4). There is some metabolic interconversion of imino compounds within a given plant, but individual concentrations remain remarkably stable and usually range between 0.1-0.5% dry weight of the plant. The compounds occur in similar concentration in seeds.

Seeds also contain a group of rare nonprotein sulphur amino acids derived from cysteine. The major sulphur compound is S-(β-carboxyethyl)-cysteine (S-CEC) which often accounts for up to 3% of the total plant dry weight (Figure 1). Lesser amounts of S-(β-carboxyisopropyl)-cysteine, djenkolic acid and N-acetyldjenkolic acid are also present. The sulphur amino acids are metabolized upon germination but persist in newly emerged leaves and sap for ten weeks or more (5). In Zapoteca, a related genus endemic to Mexico and formerly part of Calliandra, the sulphur amino acids are metabolized to volatile compounds which are exuded from the roots.

Objectives of Research

Because Calliandra is a tropical legume with potential for economic exploitation, because it is unusual in synthesizing two different groups of rare nonprotein amino acids, and because, relative to associated plants, many species suffer minimal insect predation, it has been the focus of research in the laboratory of the senior author for several years. Using a variety of techniques and bioassays, experiments have been conducted to assess what ecological significance the amino acids have as feeding deterrents and potential insecticidal agents.

Feeding Experiments - Lepidopterans

Calliandra leaf powder incorporated into artificial pinto bean based agar diets at 2.5, 5, and 10% levels and fed to Spodoptera frugiperda (fall armyworm) produces a variety of effects on all parts of the insect life cycle (6). Significant mortality occurs in the larval stage. Of those insects which survive, the larval weights are reduced and there is a delay in pupation of several days. Pupal and emerging adult weights are also reduced as is the percentage of adult moths emerging. There are no apparent malformed larvae or pupae. To some extent older larvae are able to make up early weight loss.

The nonprotein amino acids of Calliandra are moderately toxic in similar feeding tests. When an aqueous extract of leaves, from which all compounds other than the amino acids have been removed, is

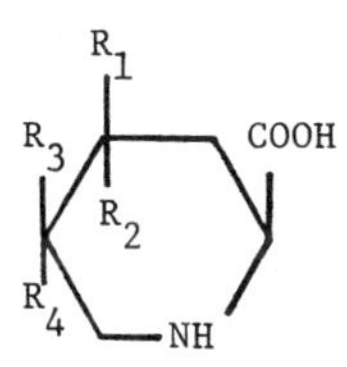

Pipecolic acid; $R_1 = R_2 = R_3 = R_4 = H$

Cis-4-Hydroxypipecolic acid; $R_1 = OH$, $R_2 = R_3 = R_4 = H$

Trans-4-Hydroxypipecolic acid; $R_2 = OH$, $R_1 = R_3 = R_4 = H$

Cis-5-Hydroxypipecolic acid; $R_3 = OH$, $R_1 = R_2 = R_4 = H$

Trans-5-Hydroxypipecolic acid; $R_4 = OH$, $R_1 = R_2 = R_3 = H$

Trans-_Trans_-4,5-Dihydroxypipecolic acid; $R_2 = R_3 = OH$, $R_1 = R_4 = H$

Trans-_Cis_-4,5-Dihydroxypipecolic acid; $R_2 = R_4 = OH$, $R_1 = R_3 = H$

Cis-_Cis_-4,5-Dihydroxypipecolic acid; $R_1 = R_3 = OH$, $R_2 = R_4 = H$

Cis-_Trans_-4,5-Dihydroxypipecolic acid; $R_1 = R_4 = OH$, $R_2 = R_3 = H$

Acetylaminopipecolic acid; $R_2 = NHCOCH_3$, $R_1 = R_3 = R_4 = H$

S-(β-Carboxyethyl)-cysteine

$$HOOC\text{-}CH_2\text{-}CH_2\text{-}S\text{-}CH_2\text{-}\underset{\displaystyle NH_2}{\underset{|}{C}}H\text{-}COOH$$

Figure 1. Pipecolic Acid Derivatives

incorporated into agar based diet at 100% plant equivalency, the growth of Spodoptera frugiperda is reduced by 20% (7). Incorporating individual amino acids into the diet produces effects similar to those of extracts from ground leaves, however, the results are variable and one cannot predict a priori which compound is likely to be most toxic (Table I). At levels at which they exist in plants (0.1-0.5%) the relatively common trans-5-hydroxypipecolic acid is more toxic to S. frugiperda than either the rare trans-trans-4,5,dihydroxypipecolic acid or acetylaminopipecolic acid. The latter compounds exhibit toxicity only at higher levels of 1 to 2.5%. The toxicity of trans-cis-4,5 dihydroxypipecolic acid is expressed at 0.5%, the upper level of its concentration in plants.

In other experiments using first stadium larvae of Spodoptera littoralis, Heliothis virescens and Heliothis armigera, individual imino compounds and S-CEC were incorporated at 0.5% into bean based diets to look for mortality effects (Table II). Larvae of S. littoralis were more susceptible to imino acids than the larvae of Heliothis. Three imino acids and the sulphur amino acid were effective. S-CEC was the only amino acid to cause significant mortality to H. armigera.

The effect of imino acids on the feeding behavior of final stadium larvae of S. littoralis, H. virescens, and H. armigera was assessed by presenting the compounds in combination with the phagostimulant sucrose on glass fiber discs. Larvae, 24-48 hours into the final stadium, were deprived of food for 4 hours and placed individually in Petri dishes containing a control and treatment disc. Each disc contained sucrose at a concentration of 0.01% w/w. The treatment discs were treated additionally with the test solution at 100 ppm. Discs were weighed at the beginning and termination of the experiment. The bioassay was terminated when 50% of any one disc was eaten (usually after 8-12 hours). An Antifeedant Index [(C-T)/(C+T)%], where C is weight of control and T of treatment disc, was calculated on the amounts eaten. Values in this Index range from +100, a potent antifeedant, to -100 a phagostimulant. Table III shows that of the compounds tested only cis-4-hydroxypipecolic acid (C4) and S-CEC significantly decreased feeding, and did so in all three species. The other compounds were either inactive (i.e. CT) or actually had some phagostimulant activity (i.e. C5 and CC).

The behavior of S. littoralis was studied in more detail during the first 2 hours of exposure to the discs by using a Video camera and measuring the following parameters: time to contact of the treatment disc, duration of this contact, and amount of time spent feeding on the treatment and control discs. S-CEC was the only compound to significantly decrease the amount of time spent feeding during the initial period of the antifeedant assay. Insects spent 40% of the time on control discs vs 11% on treatment discs. The antifeedant activity of this compound was maintained during the rest of the assay and this could have been a factor contributing to mortality in the developmental experiments (Table II). In contrast, many of the imino compounds (C4, C5, T4, and CC) actually stimulated feeding during the first two hours. In the case of C4, however, this stimulation was later reversed as shown by the Antifeedant Index. By the end of the experiment, significantly more of the control than treatment disc was consumed (Table III).

Table I. Relative Toxicities of Nonprotein Imino Acids To Freshly Hatched Spodoptera frugiperda (Adapted from Romeo, 1984)

Compound	Concentration	Mortality	Growth Inhibition
T5	0.1%	60%*	Yes
TC	0.5%	30%*	Yes
TC	1.0%	60%*	Yes
AAP	1.0%	55%*	Yes
TT	1.0%	10%*	Yes
TT	2.5%	100%*	---

n = 20; T5 = trans-5-hydroxypipecolic acid; TC = trans-cis-4,5-dihydroxypipecolic acid; AAP = acetylaminopipecolic acid; TT = trans-trans-4,5-dihydroxypipecolic acid.

*Significantly different from control (Tukey's test, $p < 0.05$).

Table II. Percent Mortality of First Stadium Larvae of Spodoptera and Heliothis Exposed to 0.5% Concentrations of Amino Acids

Compound	S. littoralis	H. virescens	H. armigera
C4	34*	14	16
C5	34*	10	6
T5	37*	24	21
CC	12	10	6
TC	15	12	10
TT	0	0	0
S-CEC	39*	21	30*

n = 20; C4 = cis-4-hydroxypipecolic acid; C5 = cis-5-hydroxypipecolic acid; T5 = trans-5-hydroxypipecolic acid; CC = cis-cis-4,5-dihydroxypipecolic acid; TC = trans-cis-4,5-dihydroxypipecolic acid; TT = trans-trans-4,5-dihydroxypipecolic acid; S-CEC = S-(β-carboxyethyl)-cysteine.

* = significantly different from control (Tukey's test for unequal sample sizes, $p < 0.05$)

Feeding Experiments - Aphids

The modest and varying effects observed thus far on behavior, growth and mortality do not suggest these amino acids have potential as insecticides against lepidopterans. A study done with aphids (*Aphis fabae*), however, provided some interesting findings. We used a bioassay in which both individual amino acids and combinations were administered by allowing second leaf stage bean seedlings of *Phaseolus vulgaris* (from which the leaves and cotyledons were removed and the top portions of the stems cut off at the level of the cotyledons) to absorb them from test solutions. After 24 hours of equilibration, aphids were confined on these plants by placing a filter paper cone around the stem. Their behavior was observed for two hours and mortality and fecundity were measured after 5 and 10 days. The bioassay has the advantages that compounds are delivered in a natural way and the experiment can be continued for a relatively long time. Aphid behavior was monitored by counting the number of probes, production of honeydew, movements on the plant, and movements off the plant to the confining filter paper cone. Some of the results are shown in Tables IV & V.

No individual compound or 2-way combination significantly affected feeding behavior. As solutions more closely approximated the actual combinations of amino acids encountered in *Calliandra* plants, however, deterrency became apparent. Dose dependent curves were seen consistently when the sulphur amino acid (S-CEC) was present together with one of the dihydroxypipecolic acid isomers (Table IV).

After 5 days exposure, *cis-trans*-4,5-dihydroxypipecolic acid (CT) was the only individual amino acid to significantly decrease the number of aphids on the plants. However, all the three and four way combinations, which contained both imino and the sulphur amino acid did so (Table V).

Solutions were removed after five days and replaced with water for the final five days. At the end of the ten day period, the numbers of aphids were recounted. Toxic effects intensified during this time, and delayed effects of some individual amino acids became apparent. For example, *trans-trans*-4,5-dihydroxypipecolic acid (TT), a compound which had little effect after 5 days, was significantly toxic after 10. Since we were able to detect the administered amino acids in the aphids at the conclusion of the experiment, it would seem we are measuring toxic rather than starvation effects.

Mode of Action

A search for the mechanism of activity of these compounds has proved elusive. Ingested nonprotein imino acids are not incorporated into proteins to produce skeletal abnormalities in a way analogous to canavanine, the nonprotein homologue of arginine (8). Some other mechanism must be involved. In a series of experiments with *Spodoptera frugiperda* (7) various nutritional indices were calculated according to Waldbaur (9). We found that *Calliandra* leaf material, the total amino acid fraction, and some of the individual imino acids depressed ECD (efficiency of conversion of digested food) and ECI (efficiency of conversion of ingested food)

Table III. Antifeedant Index [(C-T)/(C+T)%] mean (sem)

Compound	S. littoralis	H. virescens	H. armigera
C4	52.9 (4.25)*	38.2 (12.6)*	41.5 (14.4)*
C5	-41.9 (21.6)	-12.7 (5.6)	-13.4 (12.6)
CC	-29.5 (15.4)	-21.5 (4.6)	-13.8 (15.7)
T4	13.6 (15.6)	-	-
T5	21.6 (5.6)	-	-
TT	-15.6 (25.6)	- 2.5 (12.6)	- 8.4 (13.0)
TC	-15.6 (26.6)	-21.4 (21.6)	- 6.9 (21.6)
CT	8.1 (12.4)	8.6 (21.6)	13.6 (12.9)
S-CEC	51.4 (15.9)*	37.8 (12.6)*	35.6 (12.3)*

C4 = cis-4-hydroxypipecolic acid; C5 = cis-5-hydroxypipecolic acid; CC = cis-cis-4,5-dihydroxypipecolic acid; T4 = trans-4-hydroxypipecolic acid; T5 = trans-5-hydroxypipecolic acid; TT = trans-trans-4,5-dihydroxypipecolic acid; CT = cis-trans-4,5-dihydroxypipecolic acid; S-CEC = S-(β-carboxyethyl)-cysteine.

Concentration 100 ppm, n = 15-20

*Significant activity, $p < 0.05$ Wilcoxen matched pairs test.

Table IV. Deterrence Activity of Nonprotein Amino Acids to Aphid Feeding (Expressed as Percent Decrease of Controls)

Compound	Conc. (M)	10^{-5}	10^{-4}
C5		2.7	18.9
CT		19.4	5.6
PIP		8.3	0#
S-CEC		10.2	20.5
C5 + CT		14.7	8.8
C5 + PIP		5.9	11.7
C5 + CT + PIP		0#	3.2
C5 + CT + S-SEC		19.3	32.2*
PIP + CT + S-CEC		22.5*	35.4*
C5 + CT + PIP + S-CEC		38.5*	48.7*

C5 = cis-5-hydroxypipecolic acid; CT = cis-trans-4,5-dihydroxy-pipecolic acid; PIP = pipecolic acid; S-CEC = S-(β-carboxyethyl)-cysteine;

*Significantly different from control (Mann-Whitney test, one tailed, $p < 0.025$).

Number of aphids feeding was greater than control.

(Table VI). RCR (relative consumption rate) and AD (approximate digestibility) were affected by the leaf material but not by the amino acids. Thus the compounds behave like physiological toxins and apparently are taken up from the gut into the hemolymph where they adversely affect metabolism.

The similarity of the imino compounds to polyhydroxy alkaloids, which are known to act as sugar competitors and inhibit a variety of glycosidase enzymes in insects (10), led us to test these compounds against a number of enzymes. No activity of any of the mono or dihydroxy compounds was found (Fellows and Romeo, unpublished data). A trihydroxy pipecolic acid recently isolated from seeds of another legume, Baphia raccmosa, however, while having no effect on alpha and beta-glucosidase or mannosidase, is a specific inhibitor of human liver beta-D-glucuronidase and iduronidase (11). The recent detection of a probable isomer of trihydroxypipecolic acid in leaves of several Calliandra species is of interest in this regard (Swain and Morton, personal communication).

In an attempt to characterize the mode of action of the deterrent effect on the larvae of S. littoralis, correlations were made between the results of electrophysiological recordings from taste sensilla and the results of feeding behavior studies. Compounds were dissolved in sodium chloride electrolyte (0.05M) to give concentrations of 1000ppm, 100ppm and 10ppm. The maxillary styloconic sensilla of final stadium larvae of Spodoptera littoralis were stimulated with these solutions using a standard tip-recording technique (12). To allow comparison of the effect of a range of compounds on a single larva, the lateral and medial styloconic sensilla of each larva were stimulated for 1 second sequentially with 4-6 compounds, each at three concentrations with three replications per concentration. Successive stimulations of any one sensillum were separated by 2-3 minutes. The electrolyte, sodium chloride, was used as a control and was applied periodically to make sure the preparation was not malfunctioning. The solutions can stimulate 0 to 4 of the 4 chemosensitive neurones in each of the sensilla. The level of stimulation is measured by counting the number of action potentials elicited from the sensillum in the first second of stimulation.

The results of these assays allow us to attempt to correlate neural input with the resulting behavioral output. The neural input can be correlated with short term behavior, the duration of the first meal on the treatment disc, or with the longer term behavioral responses, the amount of time spent feeding on the treatment disc within the first two hours or the value of the Antifeedant Index. In previous studies neural input from the maxillary styloconic sensilla has correlated significantly with feeding behavior (13,14,15). Although the imino and sulphur amino acids stimulate 1-2 neurons in both the medial and lateral maxillary styloconic sensilla, this response does not correlate with either the duration of the first contact with the treatment disc, time spent feeding on the treatment disc in the first two hours, or the Antifeedant Index. This suggests that these sensilla are not responsible for any neural input that might be associated with the antifeedant activity recorded with C4 and S-CEC. What we do not know, however, is if these compounds interact with sucrose present on the discs used in the behavior assay. Interactions between allelochemics and sucrose

Table V. Toxicity of Nonprotein Amino Acids to Aphids (Expressed as Percent Decrease in Number Alive After 5 days Compared to Controls)

Compound	Conc. (M)	10^{-5}	10^{-4}
C5		0#	19.8
CT		19.3	30.2*
PIP		5.3	5.3
S-CEC		21.3	22.9
C5 + CT		16.3	9.5
C5 + PIP		3.4	14.6
C5 + CT + PIP		38.3*	37.6*
C5 + CT + S-CEC		59.4*	68.4*
PIP + C5 + S-CEC		55.6*	51.8*
PIP + CT + S-CEC		60.1*	60.1*
C5 + CT + PIP + S-CEC		68.8*	82.4*

Compound abbreviations - same as Table IV.

*Significantly different from control (Mann-Whitney test, one tailed, $p < 0.025$).

Number of aphids was greater than control.

Table VI. Effect of *Calliandra* leaf material and amino acids on nutritional physiology of *Spodoptera frugiperda* (Adapted from Shea, 1987)

Treatment	ECI	AD	ECD
Control	.23	.48	.54
2.5% *C. haematocephala*	.14*	.40*	.36*
Control	.27	.45	.66
Amino Acids (100% Plant Equivalency)	.27	.49	.57*
0.1% TT	.23*	.49	.52*
0.1% C5	.26	.45	.59
0.5% T5	.25	.49	.53*

ECI = efficiency of conversion of ingested food; AD = approximate digestibility; ECD = efficiency of conversion of digested food. TT = *trans*-*trans*-4,5-dihydroxypipecolic acid; C5 = *cis*-5-hydroxypipecolic acid; T5 = *trans*-5-hydroxypipecolic acid.

*Significantly different from control (ANOVA and Tukey's test for unequal sample size, $p < 0.05$)

have resulted in a decrease in the firing rate of both the maxillary styloconic sensilla, when compared to the response obtained from either the allelochemic or sucrose applied separately (13,14). In such cases, the decrease in neural input has been correlated with an increase in antifeedant activity. This remains to be investigated in this case.

Conclusion

The results of our aphid experiments emphasize the importance of looking at combinations of compounds when running bioassays to test for deterrent or toxic effects. Clearly with both the imino and sulphur amino acids we are looking at individual compounds of marginal insecticidal activity. In concert, however, they are shown to be considerably more effective. While such synergism is certainly not a new idea in chemical ecology (16), we perhaps need to give it renewed attention in our tests for biological activity. There are probably many such arrays of effective deterrents heretofore dismissed as unimportant.

Literature Cited

1. Calliandra: A Versatile Small Tree for the Humid Tropics; National Academy Press, Washington, DC, 1983; pp 3-7.
2. Uribe, B.; Ospina, A.; E. Forero. Programa Interciencia de Recursos Biologicos; Guadalupe LTDA, Bogota, Colombia, 1984; pp 104-107.
3. Romeo, J. T.; Swain, L. A.; Bleecker, A. B. Phytochemistry 1983, 22, 1615-1617.
4. Swain, L. A.; Romeo, J.T. Phytochemistry 1988, 27, 397-399.
5. Romeo, J. T.; Swain, L. A. J. Chem. Ecol. 1986, 12, 2089-2096.
6. Romeo, J.T. Biochem. Syst. & Ecol. 1984, 12, 293-297.
7. Shea, C. S. MS Thesis, University of South Florida, Tampa, Fl, 1987.
8. Dahlman, D. L.; Rosenthal, G. A. Comp. Biochem. Physiol. A. 1975, 51A, 33-36.
9. Waldbaur, G. P. Adv. Insect Physiol. 1968, 5, 229-288.
10. Evans, S. V.; Fellows, L. E.; Shing, T. K. M.; Fleet, G. W. J. Phytochemistry 1985, 24, 1953-1955.
11. Cenci di Bello, I.; Dorling, P.; Fellows, L.; Winchester, B. FEBS Lett. 1984, 176, 61-64.
12. Blaney, W. M. J. Exp. Biol. 1974, 60, 275-293.
13. Simmonds, M. S. J.; Blaney, W. M. In. Proc. 2nd. Int. Neem Conf. 1984, pp.163-18.
14. Blaney, W. M.; Simmonds, M. S. J.; Ley, S.U.; Katz, R. B. An. Phys. Entom. 1987, 12, 281-291.
15. Blaney, W. M.; Simmonds, M. S. J.; Ley, S. V.; Jones, P.S. Entomol. Exp. Appl. 1988, 46, 267-274.
16. Berenbaum M. In Chemically Mediated Interactions between Plants and Other Organisms; Cooper-Driver, G. A.; Swain, T.; Conn, E. E. Eds. Recent Advances in Phytochemistry Vol. 19; Plenum Press, New York, pp. 139-169.

RECEIVED November 2, 1988

Chapter 6

Recent Advances in Research on Botanical Insecticides in China

Shin-Foon Chiu

Laboratory of Insect Toxicology, South China Agricultural University, Guangzhou, China

Research in China on botanical insecticides has concentrated on species including *Azadirachta indica*, *Melia* spp., *Trypterygium* spp. and *Tephrosia vogelii* possessing insect growth regulating properties. These are active against the Lychee stinkbug, *Tessaratoma papilosa* and cabbageworm, *Pieris rapae*. *Tephrosia* extracts were especially active and when applied to cabbageworm larvae caused pupal deformities. Substantial differences in the concentration of insecticidal principles was found in Chinese ecotypes of *Melia* spp.. New potent insecticidal phytochemicals containing an agarofuran nucleus have been identified from species of Celastraceae. The advantages of pest control systems using botanical pesticides is discussed with reference to the problem of resistance and non-target insects.

During the last two decades, problems of rapidly increasing costs of modern synthetic organic insecticides, pest resistance to insecticides, pest resurgence and detrimental effects on non-target organisms and environmental quality, all dictate the need for effective, economical, and safe insecticides. The search for environmentally sound methods for controlling insect pests has been carried out in our laboratory for more than ten years (1,2). In this paper some highlights of recent investigations in China are presented, including experiments with plants possessing insect growth regulating properties, studies on ecotypes of the china-berry tree, *Melia* sp., studies on the bioactive principles of the Celastraceae, and experiments on the application of botanicals as a new approach to overcome insecticide resistance.

0097–6156/89/0387–0069$06.00/0

Experiments with Plants Possessing Insect Growth Regulating (IGR) Properties

With the aim of finding botanical insecticides possessing IGR properties, laboratory and field experiments have been carried out with three species of the Meliaceae: *Azadirachta indica* (neem), *Melia toosendan*, and *M. azedarach*, and several species from other families including *Trypterygium willordii*, *T. hypoglaucum*, and *Tephrosia vogelii* which were used against several agricultural insect pests.

Topical application of neem seed oil to the prothorax of nymphs of the lychee stink bug (*Tessaratoma papillosa*) showed that the oil is a strong ecdysis and growth inhibitor (Table I). The treated nymphs exhibited reduced growth and a darkened, wrinkled cuticle. Microscopic examination of the cuticle showed that the structure was malformed. Death occurred in succeeding instars.

Table I. The growth inhibiting properties of neem oil in nymphs of lychee stink bug *

Treatment (ul/insect)	Instar Treated	No.of Dead Nymphs /instar				Mortality (%)
		2nd	3rd	4th	5th	
0.005	2	1	18	2	0	70.0
control	2	1	1	2	1	16.6
0.25	3		4	16	5	86.0
control	3		2	0	0	6.6
0.5	4			9	14	92.0
control	4			1	0	4.0

* A total of 30 nymphs was used in each treatment. The neem oil (purified) was kindly supplied by Dr. H. Rembold, Max-Planck Institute, FRG. Its azadirachtin content was 400 ug/ml.

Pure, 1 - cinnamoyl - 3 - feruloyl - 11-hydroxymeliacarpin was isolated from seed kernels of *Melia azedarach* (supplied by Dr. W. Kraus). Topical application (50 ug/ml in acetone, 1 ul/insect) to third-instar nymphs of the lychee stink bug caused 23.3% of the nymphs to become deformed, with an only partially developed prothorax and abnormal antennae. The whole process of molting was also inhibited.

Topical application of 20% (w/v) neem oil (2 ul/larva) to fifth-instar larvae of the imported cabbageworm (*Pieris rapae*) markedly inhibited their growth and all died after pupation. Neem-seed kernel extract (AZT-VR-K, supplied by Dr. H. Schmutterer) was also found to be very biologically active. Spraying with a 0.3% (w/v) solution of the extract proved to be effective for as long as 21 days under field conditions. Laboratory bioassays showed that an acetone extract of the leaves of Tephrosia vogelii was a very potent growth inhibitor of the imported cabbageworm. Topical application of a 0.5% (w/v) extract (2 ul/larva) caused 89% of the larvae to fail to pupate normally (Figure 1) and the treated population gradually to die out. In the Philippines, it was found that the leaves of this legume contain 5% rotenone.

With the diamondback moth, a 10% acetone extract of the leaves of Tephrosia vogelii showed a remarkable growth inhibiting effect on fourth instar larvae - 86% of the larvae were malformed and died after feeding on the treated leaves for 48 hours.

Plants offer a large number of novel chemical compounds with biological activity and China is a very rich source of plant materials possessing IGR properties. In crop protection, the potential for the utilization of plants as insect growth regulators appears much more promising than their use as antifeedants. Research is now underway to formulate enriched products from plants with high levels of IGR activities, for practical application in the field. The chemistry and mechanisms of toxic action in insects of these plant products are also being studied. A further advantage of these IGR Plant products is their apparent lack of toxicity to non-target insects.

Studies on Ecotypes of Melia azedarach and M. toosendan.

For the practical utilization of plants as a source of insecticides, if the plant is distributed over a wide geographical area, it is important to investigate the ecotypes of the plant. A few papers have been published on the ecotypes of neem tree by Schmutterer and associates, as well as research workers in India, with particular reference to azadirachtin content (see Proceedings of the 2nd and 3rd International Neem Conferences). The china-berry trees, M. azedarach and M. toosendan, are widely distributed in China in regions south of the Yellow River. During the last few years, investigations have been carried out on the toxic components and insecticidal properties of these two species collected from different locations. Their content of toosendanin was analyzed by means of HPLC. Aqueous extracts were prepared from air-dried sanples of bark, fruit, and leaves. The extracts, after removal of fats with petroleum ether, were partitioned with chloroform. The chloroform layers were further purified by column chromatography. The chloroform layers and purified fractions were then analyzed by HPLC. Results showed that the content of toosendanin in the bark of M.

azedarach and in bark of M. toosendan varies geographically. The bark of M. azedarach growing in Yanhe, Guizhou Province (south-western China), and in Xixiang, Shanxi Province (north-western China) contains the highest amounts of toosendanin, 5.16 mg/g and 4.710 mg/g, respectively, whereas bark from Lechang, Guangdong Province (south China), contains only 1.26 mg/g. In M. toosendan, samples of the bark from northern and north-western China-Zhengzhou, Henan Province, and Yibin, Sichuan province, contain the highest amount, 4.22 mg/g and 3.03 mg/g, respectively. Samples from central and southern China, Hangzhou, Zhejiang Province and Lechang, Guangdong Province, provide low yields of toosendanin, 0.57 mg/g and 0.61 mg/g, respectively. However, the insecticidal potency of the plant, as shown by toxicity to the larvae of the imported cabbageworm, was found to be not directly correlated with the content of toosendanin.

Toosendanin was mainly found in the bark of chinaberry (3). None or very little was found in the leaves and fruit, but these plant materials also showed some toxicity to the larvae. Some extracts and crude products from the extracts of the bark showed stronger bioactivity than pure toosendanin (compared on the basis of the same amount of toosendanin). This indicates that besides toosendanin, the extracts and crude products may contain other bioactive substances with synergistic effects. It is interesting to note that the fruits of M. azedarach from the southern most part of China, Sanya, Hainan Island Province, possess fairly high toxicity to insects, although the content of toosendanin in the bark is quite low. A 1% chloroform extract of the fruits from Sanya and Haikou, Hainan Province, showed very high feeding repellence was fifth instar larvae of Spodoptera litura in a non-choice test. Mortalities of 90.6% and 96.8%, compared with 37.9%, were obtained when fifth instar larvae of imported cabbageworm were treated with a 1% chloroform extract of fruits of M. azedarach, from Sanya and Haikou, Hainan Province, or Zhengzhou, central China, respectively.

From the above preliminary results of our studies on samples of Melia collected from different regions, it can be concluded that there are distinct ecotypes in China, which vary in both the nature and quantity of their insecticidal constituents, Genotype, climatic conditions, and soil type all may play a role in the evolution of the ecotypes. More detailed studies on ecotypes will be of great value in the utilization of insecticidal plants for insect control.

Studies on Bioactive Principle from the Celastraceae

Plants of the Celastraceae are moderately well distributed in tropical and temperate regions of the world. In China there are 12 genera, composed of 183 species, of this family. They are trees, shrubs and climbers. There are about 30 Celastrus species widely distributed in the central and northwestern regions of china. The root bark of Chinese bittersweet, Celastrus angulatus Maxim. and C. glaucophyllus Rehd. et Wils have been used by farmers to control insect pests of vegetables for many years (4). It has also been reported that the root bark powder (and its ethanol extract) of C. angulatus was effective as an antifeedant with the adults and nymphs of Locusta migratoria manilensis, and the larvae of Pieris rapae. They also inhibit the reproduction of the maize weevil (Sitophilus zeamais) and act as an antifeedant and stomach poison, producing a narcotic effect. An ethyl ether extract incorporated in rice at 2.6 mg/g caused complete inhibition of reproduction of the weevil. This extract is fairly stable to sunlight and heat (5).

Recently a detailed study has been conducted on the bioactive principles of the root barks of Celastrus angulatus and C. glaucophyllus, in the Department of Phytochemistry, Kunming Institute of Botany, Yunnan Province. Altogether seven sesquiterpene alkaloids and four sesquiterpene esters have been isolated and indentified. Representative compounds are maytansine, triptolide, maytoline and evonine. All these compounds contain a common mother nucleus - agarofuran (6)(Figure 2).

Nine samples from this group of compounds have been bioassayed in our laboratory and some of them were found to be effective against the imported cabbageworm and the Asiatic corn borer (Ostrinia furnacalis) as antifeedants and stomach poisons. A new sesquiterpene, "celangalin", supposed to be the most active compound of the root bark of C. angulatus, was recently isolated (5).

It is interesting to note that the emulsified seed oil fromC. angulatus was reported to be a safe and effective insecticide for cucumber beetles and other insect pests of vegetable crops (7). Thus both the root bark and seed oil of Chinese bittersweet may serve as potential botanical insecticides.

Regarding the types of toxic symptoms and the modes of action of this class of natural products, it seems that they are closely related to the active principles in the root bark of thunder-god-vine (Trypterygium wilfordii) and the yellow azalea (Rhododendron molle). All these botanical insecticides possess antifeedant and stomach poison effects. They probably act on the central nervous system but not on cholinesterase activity. The treated insects undergo narcosis, revive briefly, then die.

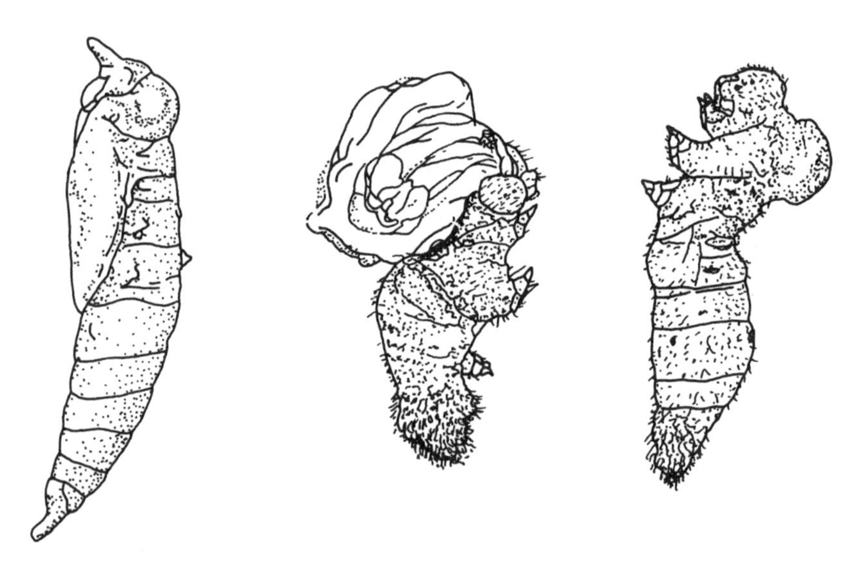

Figure 1. The effect of topical application of an acetone extract of Tephrosia vogelii on the growth and development of Pieris rapae L. Applications were made to the larval stage at rates equivalent to 2.5 ug dry weight leaves/insect (left), 25 (middle) or 100 (right).

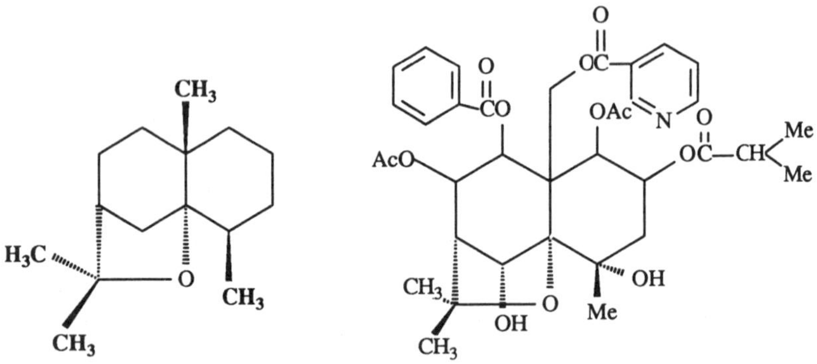

Dihydro-β-agarofuran **New compound from Celastrus angulatus**

Figure 2. Compounds from the Celastraceae.

Experiments on the Application of Plant Products and Mixtures of Botanical and Conventional Insecticides - A New Approach to Solve the Problem of Insecticide Resistance.

In chemical control of insect pests, insecticide resistance is a very serious problem. In our studies on botanicals, attempts have been made to solve this problem. One way is to make good formulations that control resistant strains of insects and also delay the occurrence of resistance in susceptible populations.

Results of laboratory and field experiments with mixtures of toosendanin, ethanolic extract of the root bark of *Trypterygium wilfordii*, and chlorpyrifos, showed marked synergistic effects in controlling the imported cabbageworm. It was found that, although conventional insecticides such as cartap and chlorpyrifos exhibit a rapid toxic action on the cabbageworm, those that survive the treatment feed and develop normally, with pupal weight similar to that of the untreated larvae. With botanical insecticides, however, either applied alone or in a mixture with other insecticides, the percentage of pupation of the larvae which survived the treatments was significantly decreased and the weight of the pupae was found to be less than that of the untreated controls.

For example, after treatment of fifth instar larvae of the imported cabbageworm with 50 ug/g crude wilforine the average weight of the pupae of the surviving individuals was 116. mg, as compared to an average weight of 162. mg of the pupae of the untreated larvae. Emergence of adults from the treated larvae was 62.5%, as compared to 93.62% for the untreated controls. The results of treatment of fourth instar larvae of the cabbageworm with a mixture of 300 ppm toosendanin and 0.5% ethanol extract of the root bark of *Trypterygium* showed that after eight days, the percentage of pupation was decreased to 0.00-7.69% as compared to the control value of 91.2%; the average weight of the pupae was 115.0 mg as compared to 183.4 for controls. The efficacy of control of the rice gall midge (*Orseolia oryzae*) was greatly increased by spraying with a mixture of a 0.5% solution of china-berry seed oil with 167 ug/ml diazinon, as compared to spraying with 500 ug/ml diazinon alone.

Because of the fact that plant derived insecticides contain a number of compounds possessing bioactive properties, insects have difficulty in adapting to them, and are not exposed to the same selection pressure as with conventional insecticides. Hence the use of botanical may avoid or delay the development of insect resistance. The mixing of conventional insecticides, such as organophosphates and carbamates, at low concentrations with botanicals can serve to increase the acute toxicity of the treatment, killing the insects more quickly. Furthermore, plant products are usually fairly selective in action; many of them have low toxicity to the natural enemies of insect pests. Field trials conducted in Guangdong in 1984 showed that a spray of 0.5% china-berry seed oil showed little toxicity to the predaceous mite, Amblyseius newsami (Erans), and other species of Amblyseius. By keeping the population density of its predator at a functional level, the damage caused by the citrus red mite (Panonychus citri McG.) might be kept below the economic threshold, and it is quite possible that the serious problem of resistance to organic and inorganic insecticides may be thus alleviated.

Our discovery of the decreases in pupal weight, percentage of emergence and egg-laying capacity of imported cabbageworm survors after treatment, with botanical insecticides, in sharp contrast to those treated with conventional organo-phosphorous compounds, may have far-reaching affects on the formation of resistant strains. There may be a genetic reason why resistant strains do not develop following treatment with natural products. It has been demonstrated that diamondback moth larvae did not develop resistance after treatment with neem extracts (AZT-VR-K) for 35 generations, whereas with deltamethrin resistance of the larvae built up within 20-35 generations. More research on the application of natural products for solving the resistance problem should be undertaken and could be very rewarding.

Plants offer a rich source of compounds possessing insect growth inhibiting properties. Besides neem, china-berry, Tephrosia vogelii, and other plant products mentioned in this paper, the potential of using phytoecdysones as IGRs must not be overlooked. It has been reported that there are more than 1000 species of plants containing these bioactive substances. In agriculture IGRs generally are compatible with the agroecosystem and may play an important role in intergrated management programs. Recently it was found that Derris extract, a time-honored botanical insecticide, also possesses IGR activity. Derris and rotenone-containing products possess many outstanding properties, one of which is their low persistance. Renewed efforts in studying this group of plants is needed.

Conclusions

Application of botanical insecticides has never given rise to the problem of resistance in agricultural insect pests. This may have biochemical, genetic and ecological implications. The rational application of these natural products either alone or in admixture with other insecticides may lead to new strategies for the prevention of the insecticide resistance problem.

For some plants of wide distribution such as neem and china-berry, investigation of their ecotypes should be based on an analysis of their main bioactive constituents as well as their biological properties. Both are essential. A detailed knowledge of the ecotypes would help us to evaluate the biosynthesis of the toxic components under different climatic conditions and to supply information on how to fully utilize the raw material locally available for insect control.

In China there is a very rich flora and farmers have been using plant products as pesticides for more than one hundred years (8), but studies on plants possessing insecticidal properties have been exploited only to a limited extent.

The bioactive principles and insecticidal action of *Celastrus angulatus* and other species belonging to Celastraceae is an example of the ample potential for research on new compounds from plants as insect control agents in China. However, international cooperation and coordination is very important in the implementation of research in this field. There is a need for multi-disciplinary international team efforts to build up and complement the valuable research currently underway at various institutions in different countries.

Literature Cited

1. Anonymous, Recent advances in the investigations of botanical insecticides. *South China Agri. Univ. Research Bull.*, 1987, No.4 19 pp Guangzhou, China.
2. Chiu, S.F. *Zeit. fur Pflanzenkr. und Pflanzenschutz*, 1985, *92*(3), 310-319.
3. Chiu, S.F.; Zhang, X. *J. South-China Agri. Univ.*, 1987, 8(2), 57-67.
4. Chiu, S.F. *J. Sci. Food Agric.* 1950, *9*, 276-286.
5. Wu, W.J.; Cao, G.J. *Acta Phytophylacica Sinica* 1985, *12*(1), 57-62.
6. Liu, J. *Investigations on the bioactive priciples of Celastraceae* 1988. Ph.D. thesis. Institute of Organic Chemistry, Lanzhou University, China.
7. Ke, Z.; Nan, U.; Lu, L.; Wu, D. *Acta Phytophylacica Sinica* 1987, *14*(3), 208-216.
8. Huang, H.T. in *Science and Civilization in China* 1987, *6*(1), 471-519.

RECEIVED December 27, 1988

Chapter 7

Antipest Secondary Metabolites from African Plants

Ahmed Hassanali and Wilber Lwande

International Centre of Insect Physiology and Ecology, P.O. Box 30772, Nairobi, Kenya

African plants used in folk medicine and in traditional pest management practices continue to yield interesting and potentially useful research leads. The leaves of the wild shrub *Ocimum suave* and the flower buds (cloves) of *Eugenia aromatica* are traditionally used as effective stored grain protectants. Eugenol, a common constituent of the two, is repellant to the maize weevil, *Sitophilus zeamais*. The medicinal plants, *Spilanthes mauritiana* and *Plumbago zeylanica* contain insecticidal isobutyl amides and the naphthoquinone, plumbagin, respectively and may have potential in small-scale mosquito control programmes. Hildecarpin, a pterocarpan from *Tephrosia hildebrandtii* is an antifeedant against the legume pod borer, *Maruca testulalis*, and the rotenoids tephrosin and rotenone are very potent antifeedants against a number of lepidopteran larvae. Comparison of the antifeedant activities of citrus limonoids and some of their modifications against the spotted stalk borer, *Chilo partellus* suggest that the units associated with antifeedant activity are spread out in the limonoid skeleton.

Plant products have played an important role in traditional medicine and in protection against pests in Africa. Chemical investigations of the plants involved have often led to the isolation of structurally and/or biologically interesting compounds. Our current concerns at ICIPE include (a) plants which may be used in crude form for small-scale protection of stored grain; (b) insecticidal plants which may have potential in small-scale mosquito control; (c) flavonoids from *Tephrosia* Pers. species (Leguminosae) for possible use in crop protection; and (d) tetranortriterpenes from local Meliaceae, Rutaceae and Simaroubaceae species and their synthetic modifications in the hope of identifying new sources of potent insect-active limonoids. The present overview highlights major accomplishments todate in these studies.

0097–6156/89/0387–0078$06.00/0

Plants used in Protecting Stored Products

Plant materials and minerals have for long been in use as traditional protectants of stored products (1). In Eastern Africa, communities in different locations appear to have evolved their own grain protection strategies based on the specific situations in their areas and the flora available around them. Unfortunately, these practices remain largely undocumented and the scientific rationale for their continued use until recent times has remained, by and large, uninvestigated.

At ICIPE, Nairobi, Kenya, we have started a modest survey of the plants used in the protection of stored grains, a simple evaluation of their efficacies and chemical investigations of their secondary components with the objective of identifying the active compounds.

Our evaluation assay consists of undamaged maize seeds (10g) in vials (3" x 3/4") to which are introduced maize weevils, *Sitophilus zeamais* (10 males and 10 females) and powdered test material (1, 2.5 and 5%). These are then compared visually with controls for 8 weeks.

In our first round of evaluation, two plant materials appeared to show protective properties: the leaves of the wild shrub *Ocimum suave* (Labiatae) and the flower buds (cloves) of *Eugenia aromatica* (Zangibaraceae). The two are used medicinally for stomach troubles and coughs, and as insect repellants (particularly against mosquitoes) and grain protectants (2,3).

Gas chromatographic and mass spectrometric examination of the essential oil of *O. suave* (Fig. 1) showed that in addition to mono and sesquiterpenes it contained eugenol (I) which has been known to be the principal volatile component of the oil of cloves (4). A gas chromatogram of oil of cloves of *E. aromatica* is shown in Fig. 2 for comparison. A literature search for biological activities of eugenol, however, revealed that the phenylpropanoid phenol had not been previously reported as an allomone of any insect. On the contrary, eugenol has been reported to be a component of an attractant blend for the Japanese beetle, *Popillia japonica* (5), and the pure compound has been demonstrated to be an attractant for the housefly (6). Our observations with *S. zeamais* weevils, however, suggested that these might be repelled by this phenolic secondary metabolite, and it became imperative to demonstrate this in a quantitatively unequivocal fashion. For this purpose, a number of Y-shaped olfactometer designs were constructed and tested. Fig. 3 illustrates the one which gave the best results. Comparison of the behaviour of the maize weevils toward different doses of eugenol and DEET (N,N-diethyltoluamide), a potent and well studied synthetic insect repellant, clearly demonstrated the repellant property of eugenol toward this insect (Fig. 4). Eugenol was found to be significantly more repellant than its isomer isoeugenol (II). Surprisingly, olfactometric tests with the oil of *O. suave* and the oil minus eugenol suggested that both the terpenoid and the phenolic fractions contributed to the repellant activity of the essential oil. The relative importance of the terpenoid components will become clear once their identification and assays on authentic samples are completed. In view of its milder (and some say more

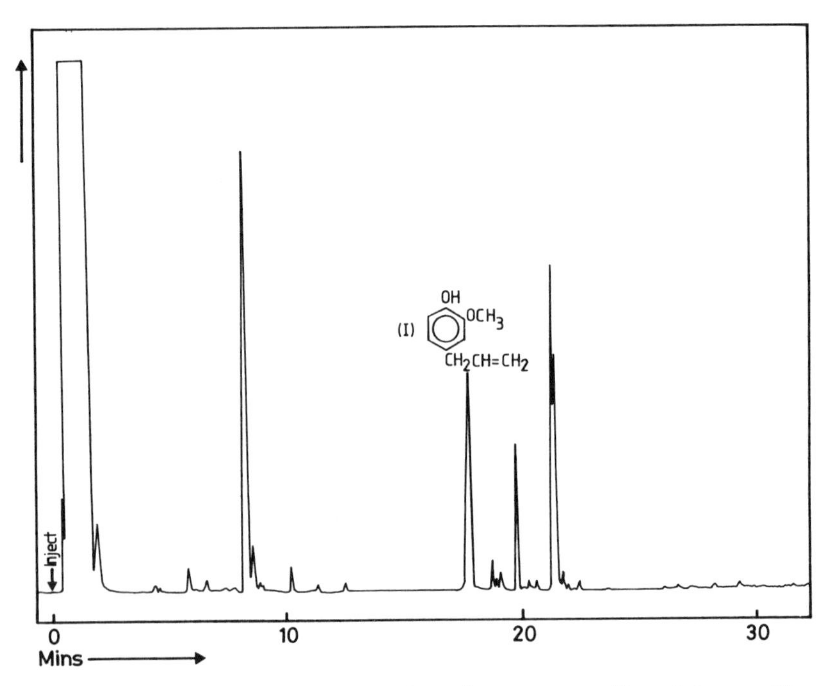

Figure 1. Gas chromatogram of *Ocimum suave* oil. Column: 15m crosslinked methyl silicone capillary column; temperature programme: 50°C (5 min) to 270°C at 50/min.

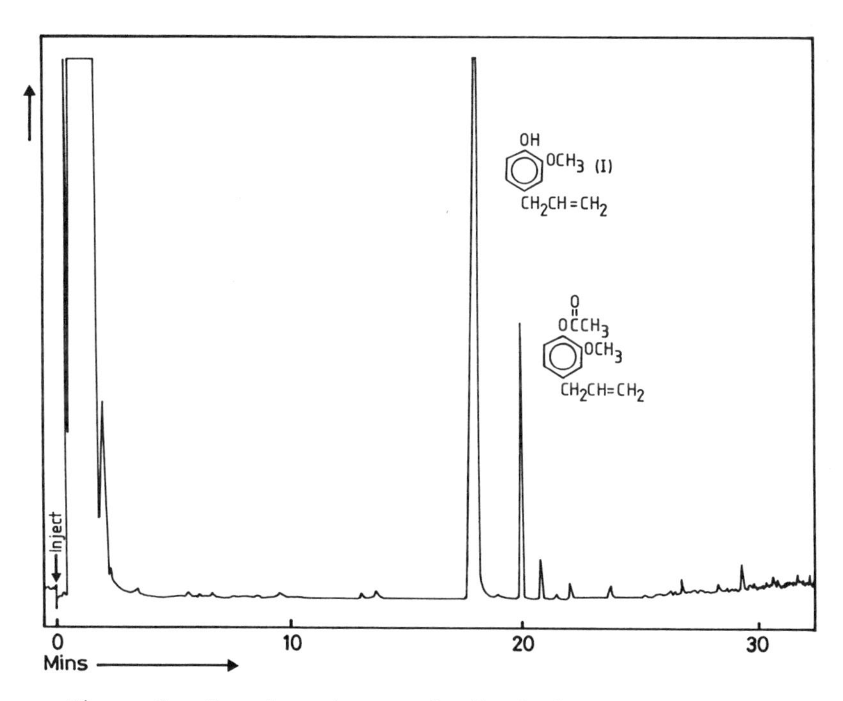

Figure 2. Gas chromatogram of oil of cloves. Column: 15m crosslinked methyl silicone capillary column; temperature programme: 50°C (5 min) to 270°C at 50/min.

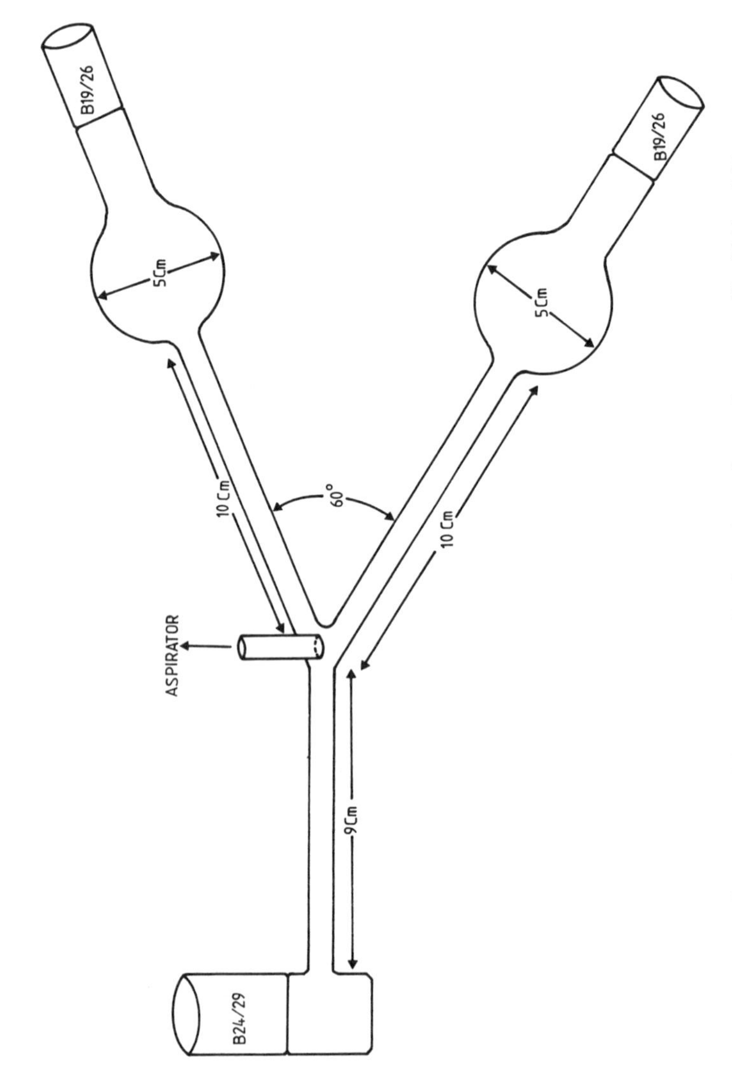

Figure 3. Olfactometer design for the maize weevil (*Sitophilus zeamais*) repellancy bioassays.

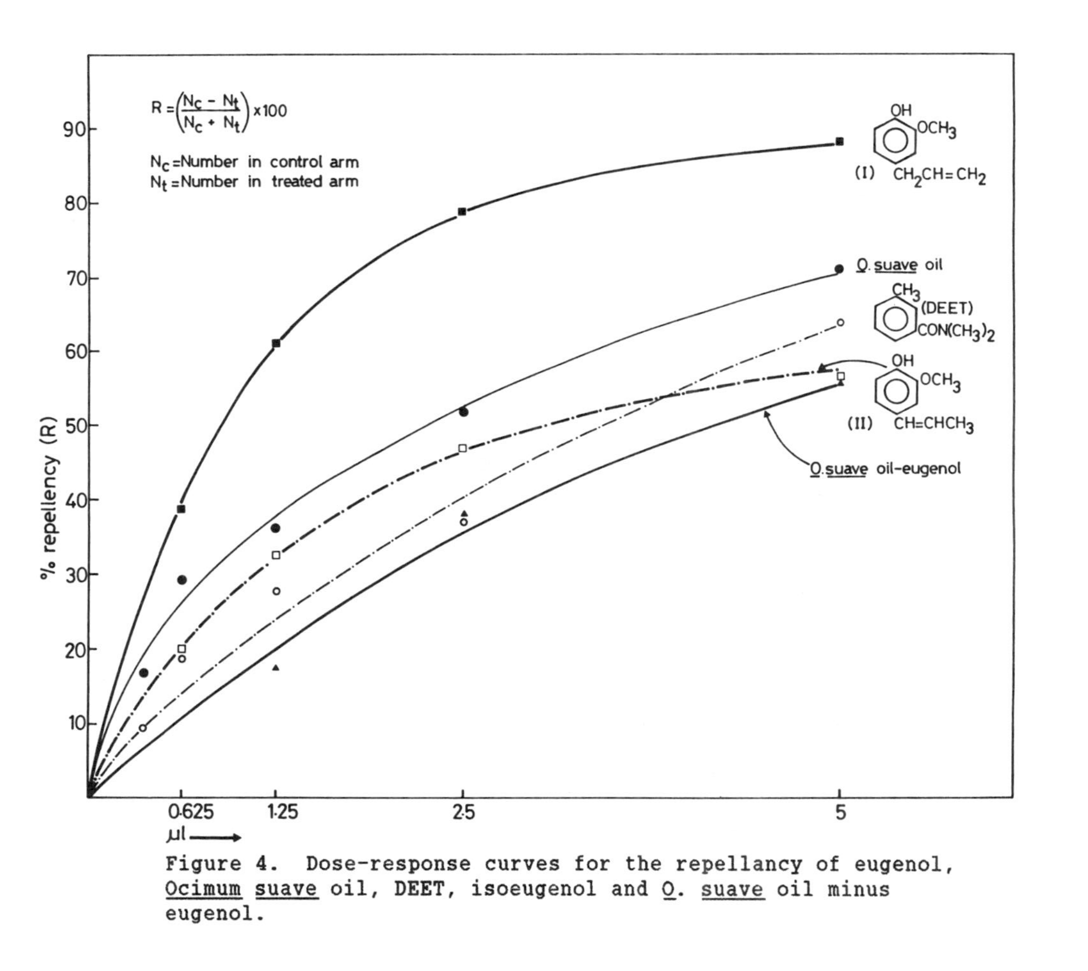

Figure 4. Dose-response curves for the repellancy of eugenol, Ocimum suave oil, DEET, isoeugenol and O. suave oil minus eugenol.

pleasant) odour than cloves and the fact that it is found thriving under varying growth conditions, *O. suave* (and other *Ocimum* species) is now our major focus for evaluation as a small-scale protectant of stored products in rural homes.

Insecticidal Plants with Potential for Mosquito Control

The breeding habits of some of the more important mosquito species in tropical Africa may provide a means for small-scale control strategies based on insecticidal plants grown by rural communities themselves. *Anopheles gambia*, the most important vector of malaria, *Culex quinquefasciatus*, the vector of Bancroftian filariasis and *Aedes aegypti*, the vector of yellow fever, all breed in small collections of water such as temporary rain puddles, man-made containers, drains, and so on (7) where the possibility exists of considerably reducing the multiplication of the mosquitoes by periodic treatment with materials derived from such plants.

Two plants investigated recently at the ICIPE, and species related to these plants, may provide candidates for such an approach to mosquito control. The first is *Spilanthes mauritiana* (Compositae), a medicinal plant used traditionally for mouth infections, stomach-ache, diarrhoea and tooth-ache (8). A methanolic extract of the wet vegetative aerial parts of the plant gave, after repeated chromatographic separations, a larvicide which was identified as dodeca-(E,E,E,Z)-2,4,8,10-tetraene N-isobutylamide (III) (9). The amide caused 100% mortality of third instar larvae of *Aedes aegypti* at 10^{-5}mg/ml (9). The chloroform extract of the flower heads of the plant gave eight additional amides (IV-XI) which are shown in Fig. 5 (10). The insecticidal properties of substituted amides of unsaturated fatty acids have been recognized for many years (11), and recently they have been a subject for more systematic structure-activity studies (12). Although their instability has been a major obstacle against commercialization, this attribute may be of special advantage for small-scale use within rural human habitats.

The second medicinal plant we have examined for mosquito larvicidal components is *Plumbago zeylanica* (Plumbiginaceae) used externally for skin disorders and internally for hookworm (13). The major active larvicidal compound was found to be plumbagin (XII) which was obtained pure from the roots of the plant in 0.15% yield. Plumbagin has previously been shown to be an antifeedant for the African armyworm *Spodoptera exempta* (14), and an ecdysis inhibitor for a number of lepidopteran larvae (15). Naphthoquinones as a group have been the subject of intensive mechanistic studies as feeding deterrents of insects (16-19). However, there has been no previous reports of their toxic effects against mosquito larvae. We have measured the larvicidal effects of a number of available structural variants of plumbagin including juglone (XIII), 2-methyl-1,4-naphthoquinone (XIV), 1,4-naphthoquinone (XV), 2,3-epoxy-1,4-naphthoquinone (XVI), 2-hydroxy-1,4-naphthoquinone (XVII), 1,4-benzoquinone (XVIII) and 1,2-naphthoquinone (XIX). Table I gives LC_{50} values of the quinones. The high activity of the parent 1,4-naphthoquinone (XV) relative to the 1,2-analogue (XIX) and 1,4-benzoquinone (XVIII), shows that the activity is largely associated with the 1,4-naphthoquinone nucleus.

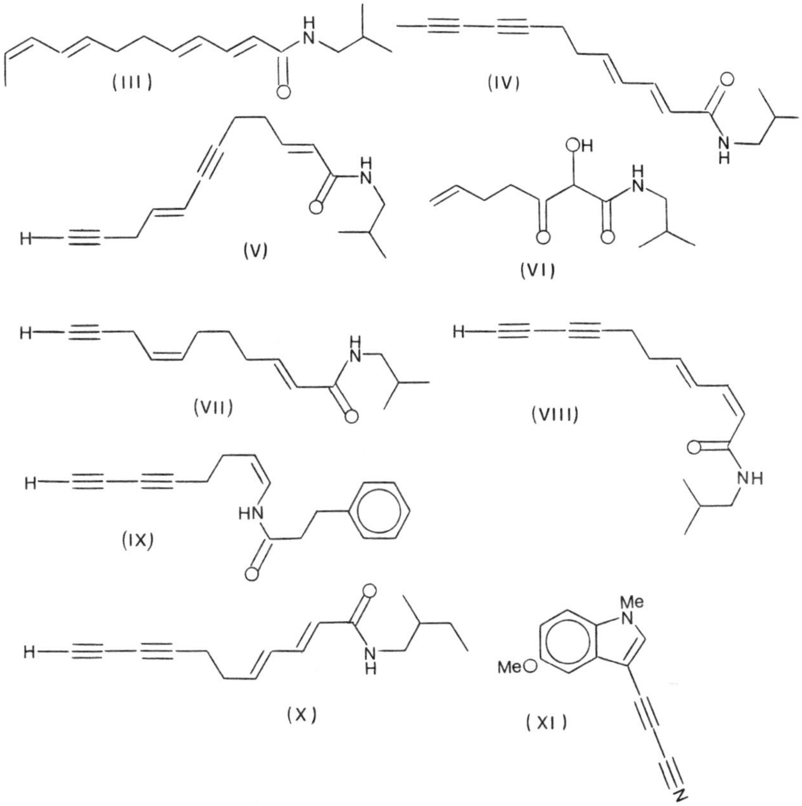

Figure 5. Structures of mosquito larvicidal amides from Spilanthes mauritiana.

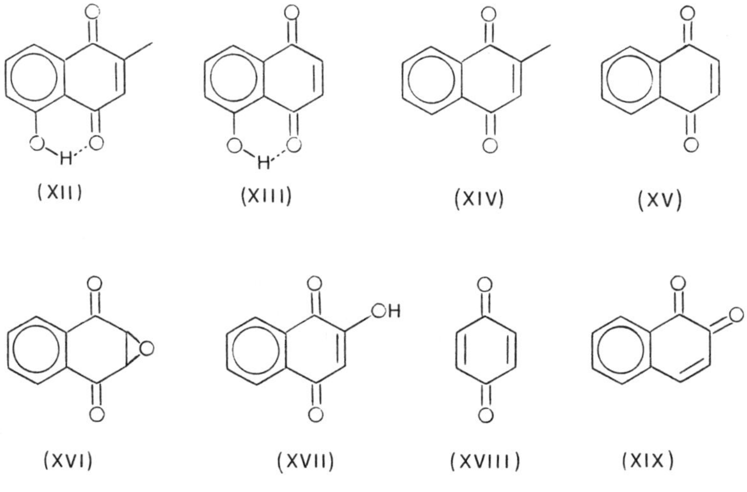

Table I. LC_{50} for various quinones against 3rd instar *Aedes aegypti* larvae*

Compound	LC_{50} (μM/20ml)
Plumbagin	0.60 ± 0.05
Juglone	0.70 ± 0.05
2-Methyl-1,4-naphthoquinone	1.15 ± 0.04
1,4-Naphthoquinone	1.82 ± 0.05
2,3-Epoxy-1,4-naphthoquinone	2.20 ± 0.05
2-Hydroxy-1,4-naphthoquinone	13.1 ± 0.2
1,4-Benzoquinone	16 ± 1
1,2-Naphthoquinone	17 ± 1

* Results obtained from 9 replicates of 20 larvae at each of 5-6 concentrations in the range 0-4 μM/20ml for the top four compounds, and 0-16 μM/20ml for the bottom four compounds.

Substitution of the 2-position of this nucleus, however, appears to lead to reduced activity with the OH substituent causing a significantly large drop. Whether this drop is associated with all the substituents in this position irrespective of their electronic effects, or specifically, with electron-donating +M groups will become clear when other substituted naphthoquinones are assayed. Naphthoquinones occur widely in tropical plants (20) and assays of a broader range of naturally-occurring structural variants may lead to the identification of readily grown plants suitable for small-scale mosquito control programmes.

Studies on Some *Tephrosia* Species

Tephrosia Pers. is a large genus of perennial and woody herbs (over 300) that are distributed in the tropical and sub-tropical regions of the world (21-22). Extracts of a number of *Tephrosia* species have been used medicinally to treat a range of ailments, in pest control as insecticides and rat poisons, and as fish poisons (22,23,24). The insecticidal and toxic effects of extracts of some *Tephrosia* species have been shown to be due to the presence of rotenoids (24,25,26). However, the majority of *Tephrosia* species that have been investigated contain non-insecticidal prenylated flavonoids (27) which have not been screened for other biological activities. In Eastern and Southern Africa, *T. vogelli* Hook f., *T. densiflora* Hook f. and *T. candida* (Roxb.) DC have been cultivated, in a limited way, by peasant farmers for use in crop protection (21,24) but, surprisingly, these practices have not developed into a significant commercial activity. Our interest in *Tephrosia* spp. is twofold: (a) to screen for rotenoid-bearing species with the purpose of generating a sound information base for their more widespread use and for the growth of an agrochemical industry; and (b) to isolate and characterize novel compounds that may be present and to assay these for anti-insect effects.

Species investigated todate include *T. hildebrandtii* Vatke and

T. elata Deflers (28,31). A number of flavonoids including some structurally novel compounds have been isolated and characterized. The anti-insect activities of two sets of compounds are of special interest. First, the pterocarpan hildecarpin (XX) showed relatively high antifeedant activity to the larvae of the legume pod-borer *Maruca testulalis* (Table II), an important pest which limits production of the cowpea crop in the tropics (32).
It bears close structural resemblance to the pterocarpan phytoalexins, medicarpin (XXI), phaseollin (XXII) and pheseollidin (XXIII), isolated from cowpea plants when infected with micro-organisms (33,35). If these phytoalexins also demostrate high antifeedant activities against *M. testulalis*, then they might constitute an interesting basis for resistance induction in the plant against this insect (28,36). Secondly, rotenoids like tephrosin (XXIV) and rotenone (XXV) are very potent antifeedants against lepidopteran African pests such as *Spodoptera exempta*, *Eldana saccharina* and *M. testulalis* (Table III), when assayed in choice bioassays using discs derived from host leaves (31). Indeed, in our assays the activities of tephrosin against *S. exempta* and *E. saccharina*, and that of rotenone against the latter are unmatched even by azadirachtin and other limonoids. Rotenoids have previously been evaluated largely as insect toxicants and doses used have been those required to cause high levels of mortality on target pests. However, in view of the present findings, a re-evaluation of different members of this group of anti-insect natural products and the manner of their use is clearly warranted.

Table II. The Feeding Inhibitory activity of Hildecarpin against *M. testulalis* Larvae

Dose (μg/disc)	Batch Number*	Control (C)	Treated (T)	Deterrence** (D) %	
100	1	11.86	1.97	83.4)	
	2	29.19	2.07	92.9)	85.8 ± 6
	3	61.58	11.69	81.0)	
10	4	28.27	13.62	51.8)	
				)	51.9 ± 0.1
	5	52.91	25.41	52.0)	

* For each batch, 20 larvae of the test insect were used. Two larvae were allowed to feed on a pair of treated disc (T) and control disc (C). The result of each batch is based on a summation of 10 such feeding assays. **Defined as (1 - T/C) x 100. (Reproduced with permission from Ref. 28. Copyright 1985 Pergamon Press).

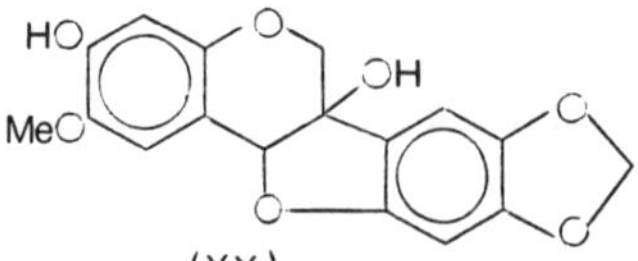

HO
MeO
OH
(XX)

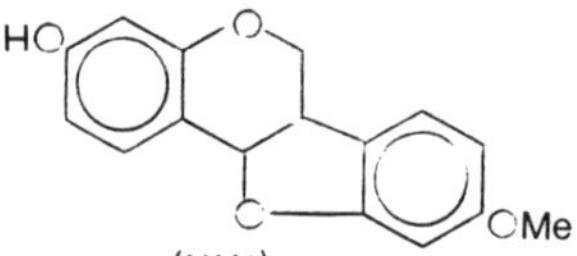

HO
OMe
(XXI)

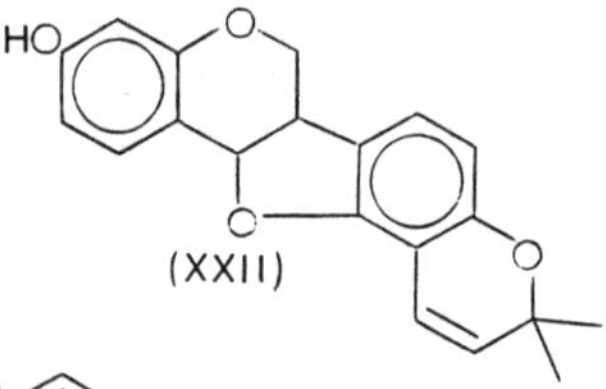

HO
(XXII)

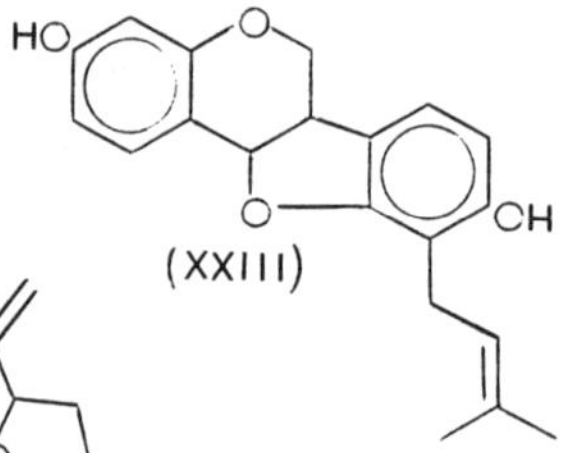

HO
OH
(XXIII)

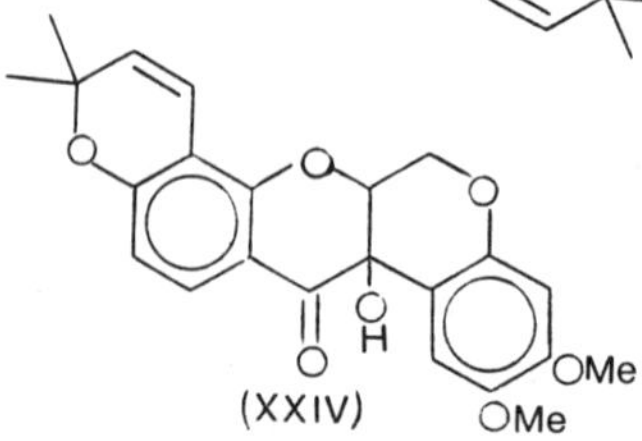

OMe
OMe
H
(XXIV)

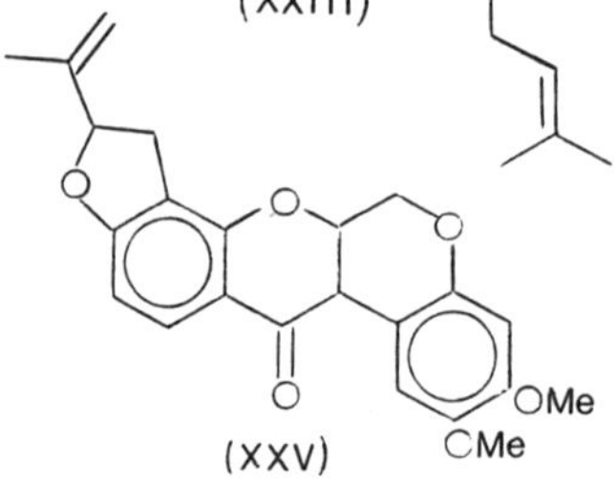

OMe
OMe
(XXV)

Table III. Antifeedant activities of some rotenoids against some African insect pests

Compound μg/disc	*Spodoptera exempta*	*Eldana saccharina*	*Maruca testulalis*
Tephrosin			
100	*	94 ± 6(3)	
10	89 ± 8(5)	86 ± 5(3)	
1	89 ± 5(5)	72 ± 7(3)	
Rotenone			
100	58 + 18(3)*	89 ± 2(4)	97 ± 2(4)
10	82 ± 7(3)	92 ± 5(3)	86 ± 8(4)
1	52 + 4(3)	81 + 11(3)	45 + 11(3)

* Evidence of toxication, lithargy, regurgitation and much reduced discrimination between treated and untreated discs. (Reproduced with permission from Ref. 31. Copyright 1987 ICIPE Science Press).

Limonoids and their Analogues

Recent isolation of novel anti-insect tetranortriterpenes belonging to the genera other than *Melia*, such as *Trichilia* (37) and *Turreae* (Rajab, M. et al., Phytochemistry, in press) and to families other than Meliaceae such as Simaroubaceae (38,39,40), attest to the existence of rich, yet undiscovered sources of this class of compounds in African plants belonging to several families and genera. The major objectives of our research on limonoids are to identify limonoid-bearing plants that thrive in different African environments and to elucidate the structural requirements for the antifeedant and growth-disruptive effects by structure-activity studies of the natural limonoids and their synthetic modifications against the African crop pests.

The present work, which has involved the study of antifeedant activities of citrus limonoids and their synthetic modifications, has been carried out in collaboration with Professor Michael Bentley and his co-workers at the Univeristy of Maine, U.S.A. Two different insects were chosen for the study, Colorado potato beetle, *Leptinotarsa decemlineata* and the stem-borer, *Chilo partellus*. Results on *L. decemlineata* have been reported elsewhere (Bentley, M.D. et al., Entomol. Exp. Appl., in press) and here we shall highlight our findings on *C. partellus*.

The limonoids tested include limonin (XXVI) deoxyepilimonol (XXVII), epilimonyl acetate (XXVIII), obacunone (XXIX), nomilin (XXX), limonin diosphenol (XXXI), deoxylimonin (XXXII), epilimonol (XXXIII), deoxytetrahydrolimonin (XXXIV) and hexahydrolimoninic acid (XXXV). Each of these compounds was assayed at three doses against fifth instar *C. partellus* larvae in a choice arena of treated and untreated discs of maize leaf, similar to that reported earlier (41). The results are summarised in Fig. 6. The most noteworthy feature of these results is that none of the structural alterations represented by the different limonoids lead to a total loss of antifeedant activity. This is in contrast to results obtained for

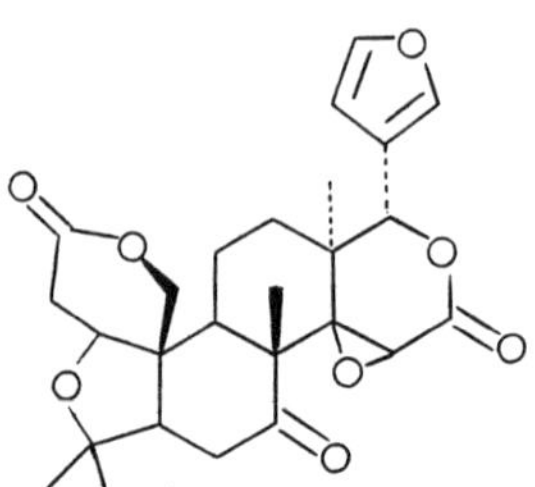

(XXVI) 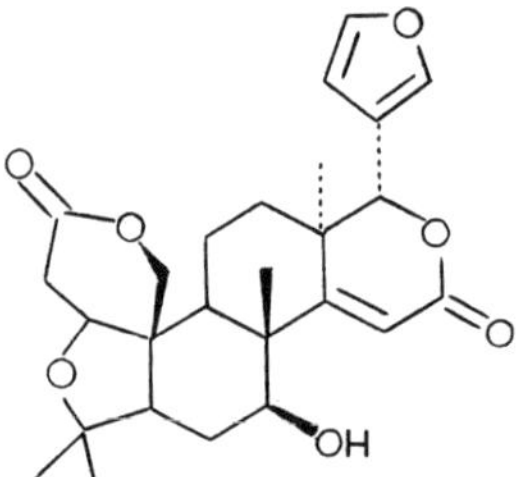 (XXVII) 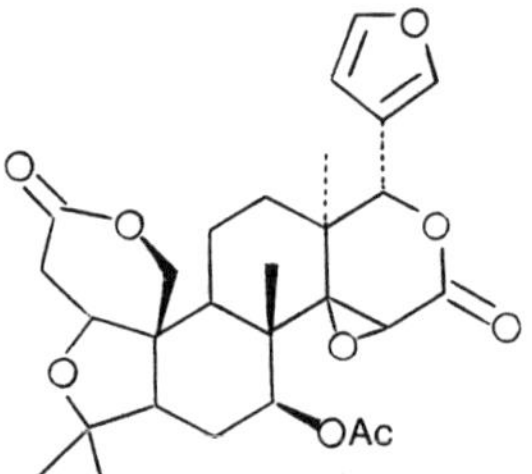 (XXVIII)

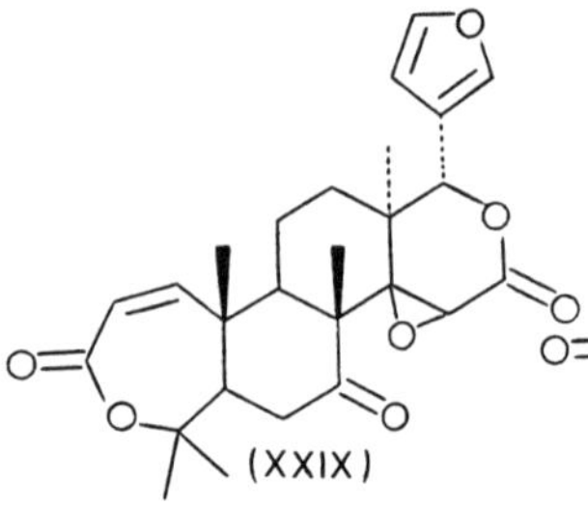

(XXIX) 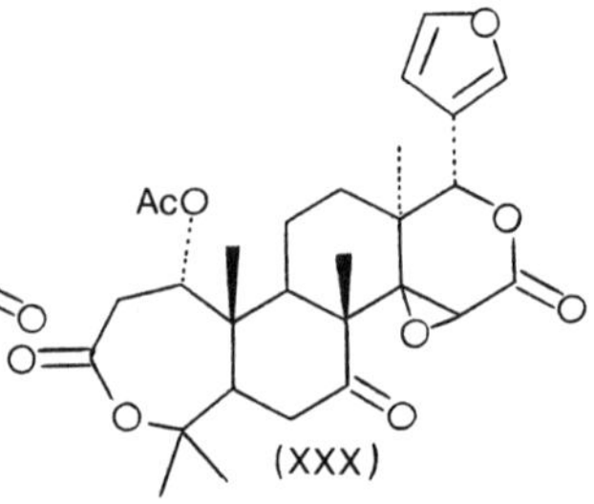 (XXX)

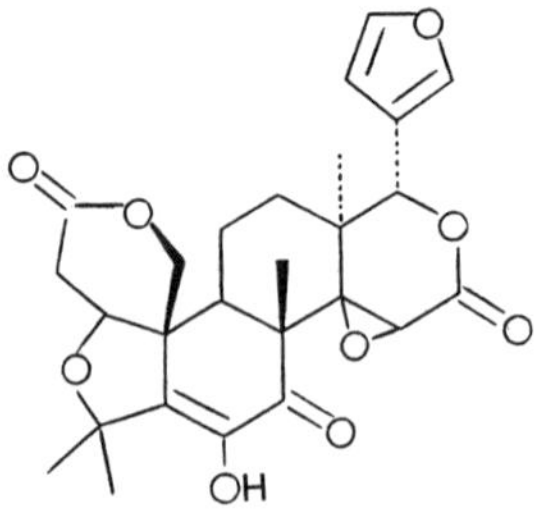

(XXXI) 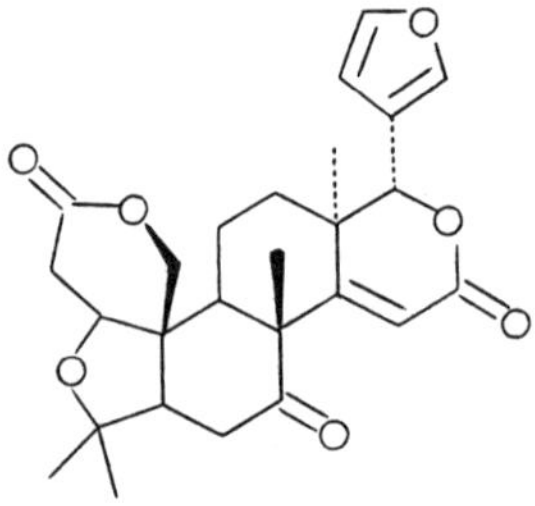 (XXXII) 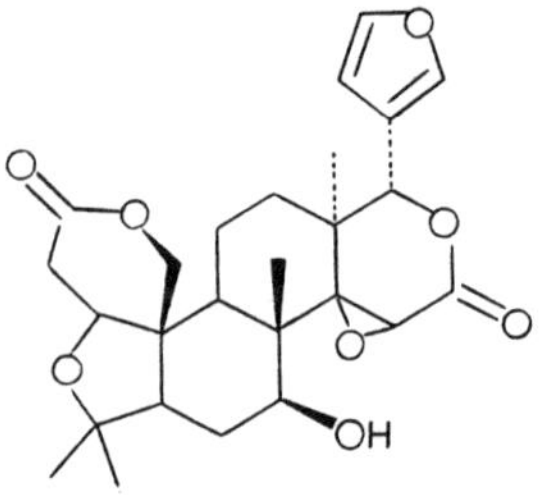 (XXXIII)

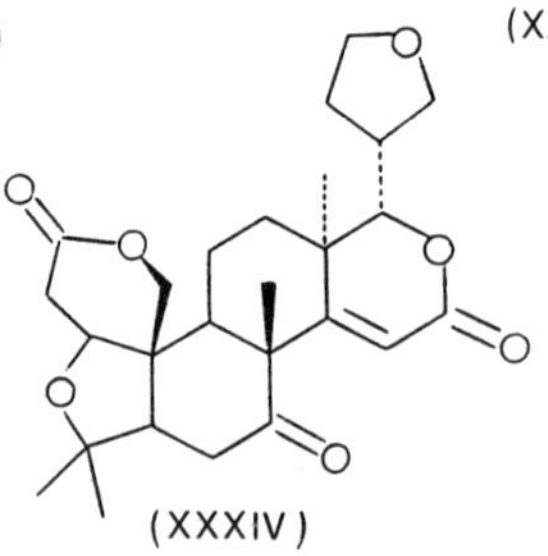

(XXXIV) 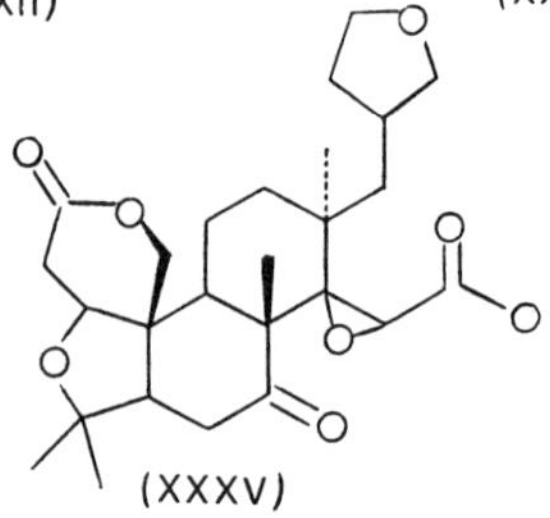 (XXXV)

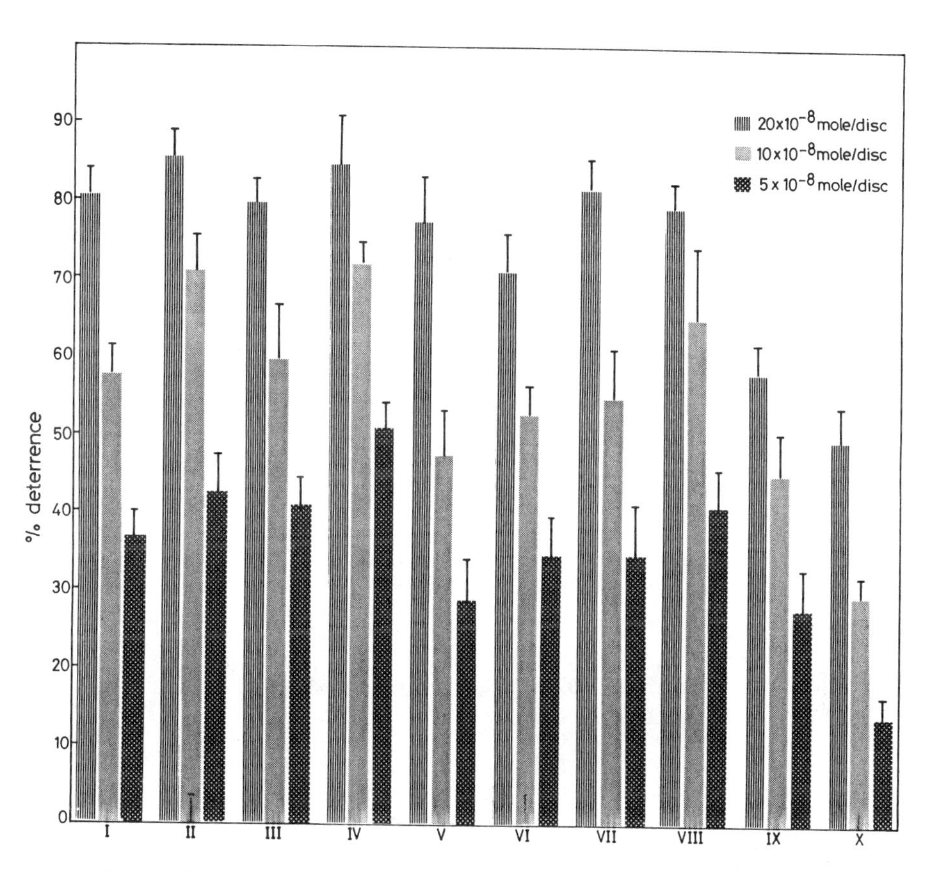

Figure 6. Antifeedant activity of modified limonoids against _Chilo partellus_ larvae.

the Colorado potato beetle where the presence of the unsaturated furan moiety appears to be crucial for significant antifeedant activity of the limonoid. Nevertheless, a point of similarity between the two insects is that the furan group is also an important requirement for high antifeedant activity against *C. partellus*. Thus deoxytetrahydrolimonin (XXXIV) is significantly less active at 20 x 10^{-8}mole/disc than deoxylimonin (XXXII), and hexahydrolimoninic acid (XXXV) is the least active in the series at all the three doses tested. Several other differences between the feeding responses of the two insects emerge from the two sets of data. Significantly higher activities of obacunone (XXIX) relative to those of limonin (XXVI) at the two lower doses, and nomilin (XXX) at all the three doses, clearly demonstrate the role of ring A in interaction with the receptor molecule in *C. partellus*. In assays with the Colorado potato beetle, however, modification of ring A of limonin produced no significant change in the antifeedant activity of the resulting compound suggesting no significant role for this end of the molecule. The epoxy group of ring D is important for high activity against the Colorado potato beetle. On the other hand, its removal produces no significant change in activity against *C. partellus*. A summation of the data on the Colorado potato beetle suggests that the structural units associated with antifeedant activity are largely localised in ring D and the furan group. In *C. partellus* these appear to be spread out in the limonoid skeleton, an inference which was also arrived at in an earlier study with two other lepidopteran African pests *E. saccharina* and *M. testulalis* (42).

Literature Cited

1. Golab, P., Webley, D.J. The use of Plants and Minerals as Traditional Protectants of Stored Products, Tropical Products Institute: London, Report G 138, 1980.
2. Kokwaro, J.O. Medicinal Plants of East Africa; East African Literature Bureau: Nairobi, 1976; p111.
3. Williams R.O. The Useful and Ornamental Plants in Zanzibar and Pemba; St. Ann's Press: Timperley, Altrincham, 1949; p383.
4. The Merck Index, Merck & Co., Inc., 1983, 10th ed.
5. Ladd, T.L. J. Econ. Entomol. 1980, 73, 718-720.
6. Sharma, R.N., Saxena, K.N. J. Med. Entomol. 1974, 11, 617-621.
7. Gillet, J.D. Common African Mosquitoes; Heinemann: London, 1972; p26,68 and 102.
8. Kokwaro, J.O. Medicinal Plants of East Africa; East African Literature Bureau: Nairobi, 1976; p71.
9. Jondiko, I.J.O. Phytochem. 1986, 25, 2289-2290.
10. Jondiko, I.J.O.; Achieng' G., Pattenden, G. Fourteenth Annual Report, ICIPE, Nairobi, 1986, p84.
11. Jacobson, M. In Naturally Occurring Insecticides; Jacobson, M.; Crosby D.G., Eds., Dekker, New York, 1971; p139-176.
12. Elliot, M. In Recent Advances in the Chemistry of Insect Control; Janes, N.F. Ed.; Royal Society of Chemistry, London, 1985; p73-102.
13. Kokwaro, J.O. Medicinal Plants of East Africa; East African Literature Bureau: Nairobi, 1976; p177.

14. Kubo, I.; Taniguchi, M.; Chapya, I.; Tsujimoto, K. Planta Med. 1980, 185-187.
15. Kubo, I.; Uchida, M.; Klocke, J.A. Agr. Biol. Chem. 1983, 47, 911-913.
16. Norris, Dale M., Bioelectrochem. Bioenerg. 1985, 14, 449-456.
17. Singer, G., Rozenthal, J.M., Norris, Dale M. Nature (London). 1975, 256, 222-3.
18. Norris, Dale M., Chu, Hsien-Ming. J. Insect Physiol. 1974, 20, 1687-1696.
19. Gilbert, B.L.; Baker, J.E.; Norris, Dale M. J. Insect Physiol. 1967, 13, 1453-1459.
20. Marini Bettolo, G.B. In Natural Products for Innovative Pest Management; Whitehead, D.L.; Bowers, W.S., Eds.; Pergamon: Oxford, 1983; p201.
21. Dale, I.R.; Greenway, P.J. Kenya Trees and Shrubs; Buchanan's Kenya Estates Ltd.: Nairobi, 1961; p654.
22. Gillett, J.B.; Polhill, R.M.; Verdcourt, B. In Flora of Tropical East Africa; Milne-Redhead, E.; Polhill, R.M., Eds; The Government Printer: Nairobi, 1971 p501.
23. Kokwaro, J.O. Medicinal Plants of East Africa; East African Literature Bureau: Nairobi, 1976; p384.
24. Mitchell Watt, J., Brandwijk, M.G. The Medicinal and Poisonous Plants of Southern and Eastern Africa; Livingstone: London, 1962; p1457.
25. Dalziel, J.M. The Useful Plants of West Tropical Africa; Crown Agents: London, 1955; p612.
26. Marini Bettolo, G.B. In Natural Products for Innovative Pest Management; Whitehead, D.L.; Bowers, W.S., Eds.; Pergamon: Oxford 1983; p190.
27. Gomes, C.M.R.; Gottlieb, O.R.; Marini Bettolo, G.B.; Delle Monache, F.; Polhill, R.M. Biochem. System. Ecol. 1981, 129-147.
28. Lwande, W.; Hassanali, A.; Njoroge, P.W.; Bentley, M.D.; Delle Monache F.; Jondiko, J.I. Insect Sci. Applic. 1985, 6, 537-541.
29. Delle Monache, F.; Labbiento, L.; Marta, M.; Lwande, W. Phytochemistry 1986, 25, 1711-1713.
30. Lwande, W.; Bentley, M.D.; Macfoy, C.; Lugemwa, F.N.; Hassanali, A; Nyandat, E. Phytochemistry 1987, 26, 2425-2426.
31. Bentley, M.D.; Hassanali, A.; Lwande, W.; Njoroge, P.E.W.; Ole Sitayo, E.N.; Yatagai, M. Insect Sci. Applic. 1987, 8, 85-88.
32. Singh, S.R.; Emden, H.F. Annu. Rev. Ent. 1979, 24, 255-278.
33. Lampard, J.F. Phytochemistry 1974, 13, 291-292.
34. Bailey, J.A. J. Gen. Microbiol. 1973, 75, 119-123.
35. Preston, N.W. Phytochemistry 1975, 14, 1131-1132.
36. Kogan, M.; Paxton, J. In Plant Resistance to Insects; Hedin, P.A., Ed.; ACS Symposium Series No. 208; American Chemical Society: Washington, DC, 1983; p153-171.
37. Nakatani, M.; James, J.C.; Nakanishi, K. J. Am. Chem. Sec. 1981, 103, 1228-1230.
38. Kubo I.; Tanis S.P.; Lee Y.W.; Miura I.; Nakanishi K.; Chapya A. Heterocycles 1976, 5, 485-497.
39. Liu, H-W; Kubo, I.; Nakanishi, K. Heterocycles 1982, 17, 67-71.

40. Hassanali, A.; Bentley, M.D.; Slawin, A.M.Z.; Williams, D.J.; Shepherd, R.N.; Chapya, A. Phytochemistry 1987, 26, 573-575.
41. Hassanali, A; Bentley, M.D.; Ole Sitayo, E.N.; Njoroge, P.E.W.; Yatagai, M. Insect Sci. Applic. 1986, 7, 495-499.
42. Hassanali, A.; Bentley, M.D. In Natural Pesticides from the Neem Tree and other Plants; Schmutterer, H., Ascher, K.R.S.; Eds.; Proc. 3rd Int. Neem Conf.; Gesellschaft fur Technische Zusammenarbeit (GTZ): Eschborn, 1987; p683.

RECEIVED November 2, 1988

Chapter 8

Insecticidal Activity of Phytochemicals and Extracts of the *Meliaceae*

Donald E. Champagne[1], Murray B. Isman[2], and G. H. Neil Towers[1]

[1]Department of Botany, University of British Columbia, Vancouver, British Columbia V6T 2B1, Canada
[2]Department of Plant Science, University of British Columbia, Vancouver, British Columbia V6T 2B1, Canada

Limonoids, characteristic natural products of the Meliaceae, Rutaceae, and other Rutales, have marked biological activity against a variety of insects. In particular, the well-known compound azadirachtin is under development as a possible commercial insecticide owing to its potent antifeedant and growth-regulating properties. This compound inhibits the feeding, growth and survival of the variegated cutworm, *Peridroma saucia*, with an EC_{50} and LC_{50} of 0.36 and 2.7 ppm in diet, respectively. In contrast, azadirachtin has no antifeedant activity against nymphs of the migratory grasshopper, *Melanoplus sanguinipes*, but does inhibit molting in a dose-dependant manner with an oral ED_{50} of 11.3 μg/g insect fwt. The ED_{50} via intrahemocoelic injection is 3.2 μg/g, indicating that the gut poses a physical or physiological barrier to azadirachtin bioavailability. The lesser toxicity via the oral route may be largely due to local metabolism, as the oral toxicity can be synergized by piperonyl butoxide, an inhibitor of mixed-function oxidases. The topical ED_{50} is 4.5 μg/g insect, indicating that azadirachtin is well absorbed through the integument. Azadirachtin toxicity in this insect does not involve inhibition of sterol reductases or sterol transport. Of nine other limonoids tested, only cedrelone, anthothecol, and bussein inhibit larval growth of *P. saucia*, and only cedrelone inhibits the molting of the milkweed bug, *Oncopeltus fasciatus*. Foliar extracts of species in the subfamily Melioideae are on average more insecticidal than extracts from the Swietenioideae; some deserve further attention as sources of insecticidal phytochemicals.

In recent years much effort has been devoted to the search for alternatives to the current arsenal of pest-control chemicals. Plant-derived extracts and phytochemicals, which once formed the basis of pest-control technology, are again being scrutinized for potentially useful products or as models for new classes of synthetic insecticides (*1, 2*). Amongst the most promising of the natural products investigated to date are the limonoids, characteristic

0097–6156/89/0387–0095$06.00/0

secondary metabolites in the Rutales, particularly the families Meliaceae and Rutaceae (*3, 4*).

The best known of the limonoids is azadirachtin (Figure *1*), a constituent of the leaves and fruits of the neem tree, *Azadirachta indica* A. Juss., and the closely related chinaberry, *Melia azediracht* (*5*). Azadirachtin was first identified as a feeding deterrent for the desert locust, *Schistocerca gregaria* (*6*), and has subsequently been reported to have antifeedant activity against nearly two hundred species of insects (Jacobson, this volume). Other limonoids also have antifeedant activity at higher concentrations (*3, 7-13*). In addition to its antifeedant action, in many insects azadirachtin disrupts the growth and metamorphosis of insects by interfering with the production of β-ecdysone and juvenile hormone by an as yet unknown mechanism (*14-20*). The extremely low mammalian toxicity of this compound (*21-23*), and its purported systemic action in a variety of important crop plants including corn (*24*) and rice (*25*), are further factors favoring the development of azadirachtin as a commercial insecticide.

We have investigated the bioactivity of azadirachtin against two phytophagous insects of importance to Canadian agriculture, the migratory grasshopper, *Melanoplus sanguinipes* Fab. (Orthoptera: Acrididae), and the variegated cutworm, *Peridroma saucia* (Hubner) (Lepidoptera: Noctuidae). Both are highly polyphagous herbivores but neither encounter limonoids in their natural host range. The antifeedant, growth inhibiting, and molt inhibiting activity of nine other limonoids from various Meliaceae, Rutaceae, and Simaroubaceae were assayed against *P. saucia* and the milkweed bug, *Oncopeltus fasciatus* (Hemiptera: Lygaeidae). Foliar extracts from twenty species of Meliaceae were assayed for inhibition of growth and toxicity against *P. saucia*.

Azadirachtin Toxicity to *Peridroma saucia*

Azadirachtin exhibits a remarkable degree of biological activity when fed in artificial diet to *P. saucia* larvae (*26*). When neonate larvae were reared for a seven-day period on diet spiked with azadirachtin, the concentration required to reduce growth by 50% relative to the controls (EC_{50}) was found to be 0.5 nmol/g diet fresh weight (fwt) (0.36 ppm), and the LC_{50} was 3.8 nmol/g diet (2.7 ppm) (Figure 2). Antifeedant effects of azadirachtin were investigated using a two-choice bioassay: 5 cm diameter petri dishes were marked off in quadrants, and control or azadirachtin treated (0.15, 0.5, or 1.5 nmol/g fwt) diet cubes were placed in quadrants. Ten neonate or 6-day-old (second instar) *P. saucia* were released in the center of each dish, which were put in an opaque box to eliminate phototactic effects. The numbers of larvae on each diet cube and/or in each quadrant were then counted after 24 hours. Azadirachtin concentrations of 0.5 nmol/g diet fwt or higher significantly deterred feeding by neonates (Table I). This sensitivity is apparently lost as the larvae develop, as six-day old larvae showed no significant response to azadirachtin concentrations as high as 1.5 nmol/g fwt.

The influence of azadirachtin on growth, consumption, and dietary utilization (*27*) were examined in third instar larvae. The relative growth rate (RGR) was decreased by almost 50% with 1.5 nmol/g fwt azadirachtin, whereas no significant effect was seen at 0.5 and 0.15 nmol/g fwt (Table II). This decrease was largely due to a decrease in the relative consumption rate (RCR) and secondarily to a decrease in the efficiency with which ingested (ECI) and digested (ECD) food is converted to insect biomass. These effects parallel those seen with other lepidopterans including the European corn

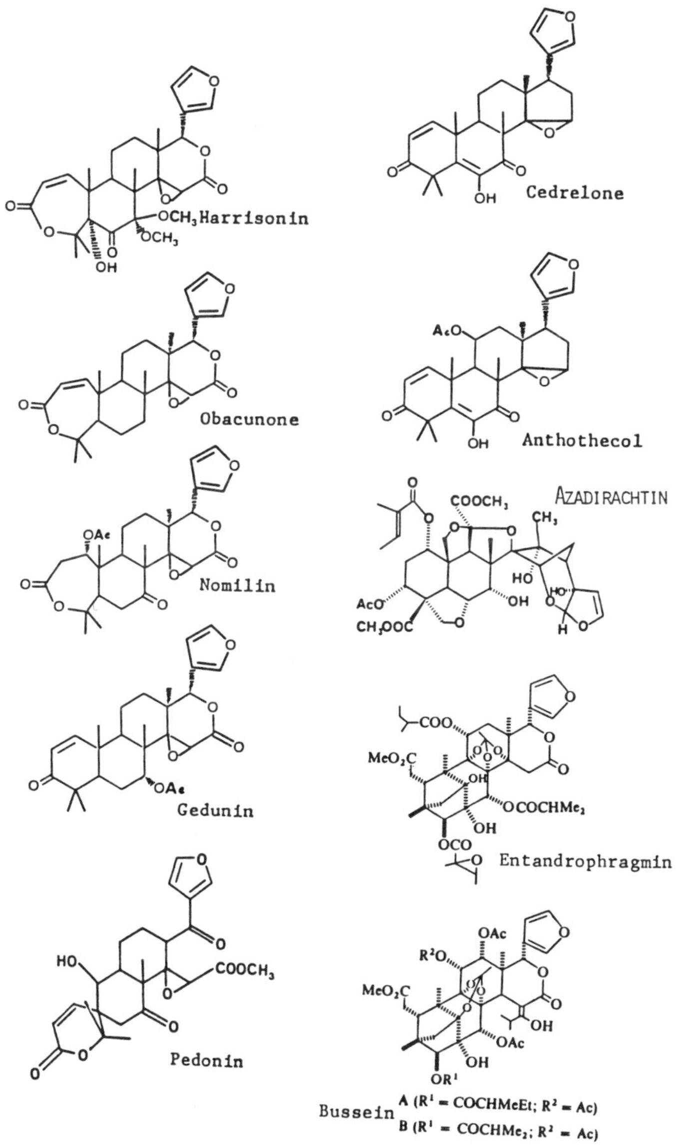

Figure 1. Structures of limonoids discussed in the text.

borer, *Ostrinia nubilalis* (*28*),the tobacco budworm, *Heliothis virescens* (*29*), and the cabbage webworm, *Crocidolomia binotalis* (*30*). The observed increase in the approximate digestibility (AD) may be due to a slower passage of the food through the gut, as has been suggested in the case of *Locusta migratoria* (*31*).

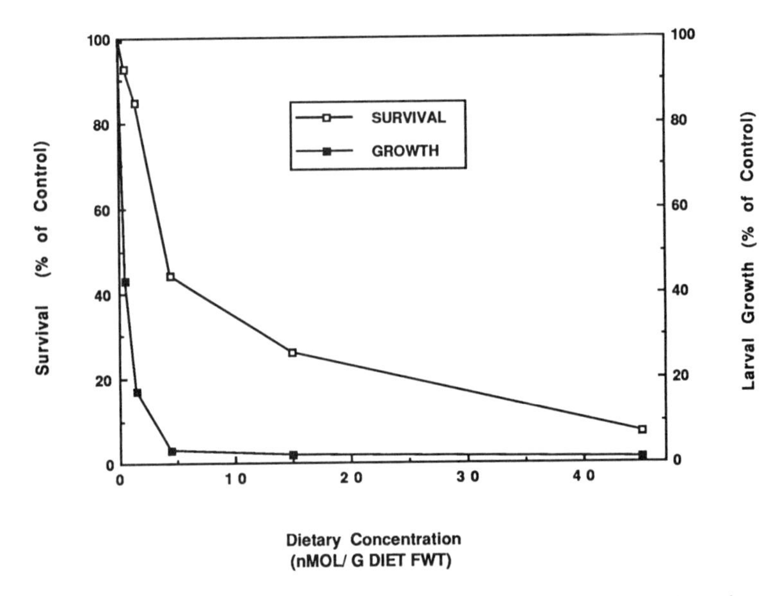

Figure 2. Effect of azadirachtin on *Peridroma saucia* growth and survivorship over a seven day assay.

Table I. Effect of azadirachtin on diet choice by neonate and 6-day-old *Peridroma saucia*. Asterisks indicate a significant difference between control and treated diet (X^2, $p < .05$)

Azadirachtin nmol/g fwt	Percentage of larvae on control (C) or treated (T) diet at 24 hr. neonate C	neonate T	6 day old C	6 day old T
0	50	50	53	47
0.15	36	64	59	41
0.5	70	30*	--	--
1.5	89	11*	47	53

Table II. Effect of azadirachtin on dietary utilization by third instar *P. saucia*. Values given are the mean + 1 standard deviation

Dose nmol/g fwt	RGR (mg/mg/d)	RCR (mg/mg/d)	AD (%)	ECI (%)	ECD (%)
Control	0.27±.03	1.12±.03	50.4± 3.0	27.6± 6.9	54.8±18.0
0.15	0.25±.01	1.03±.03	40.4± 2.6	23.7± 4.3	58.7±15.4
0.5	0.22±.05	0.51±.03	51.0±14.0	25.1±11.9	49.2±22.0
1.5	0.14±.09	0.44±.02	57.5± 6.1	12.3± 8.2	21.4±18.5

RGR is the Relative Growth Rate, RCR is the Relative Consumption Rate, AD is the Approximate Digestibility, ECI is the Efficiency of Conversion of Ingested diet, and ECD is the Efficiency of Conversion of Digested diet. All values were calculated according to Reese and Beck (1976).

Azadirachtin Effects on *Melanoplus sanguinipes*

Amongst the few insects reported to be resistant to the effects of azadirachtin are the new world grasshoppers, including the migratory grasshopper *Melanoplus sanguinipes* (*32*). Indeed we found that fifth instar nymphs of this species would readily consume cabbage leaf discs treated with up to 500 ppm azadirachtin in a single feeding bout, confirming the lack of chemosensory-based antifeedant effects against this insect. However, nymphs fed azadirachtin subsequently failed to molt or molted to severely deformed adults. This prompted us to examine the dose-response of orally-administered azadirachtin (*33*). The effects observed were divided into four categories to allow quantitation (Figure *3*). At doses of up to 8 μg/g insect fwt, most individuals were able to molt successfully (Category I), although at the higher doses molting was delayed by 3 or 4 days compared to the controls (all of which molted on day 8 or 9 of the instar). Only a small percentage of the treated nymphs molted to adults with deformed wings (Category II). At 10 μg/g most nymphs completed the molt as severely deformed adults with crumpled and curled wings and, in some cases, deformed tarsi as well (Category II). These individuals were very slow in completing the molt, and probably began to sclerotize new cuticle before ecdysis was complete. At this dose some nymphs did molt normally (Category I); a small number died in a failed molt attempt (Category III). An abrupt transition in effects was seen between 10 and 13 μg/g insect. At the lower dose, 90% of the nymphs were able to complete the molt (categories I and II combined), but at 13 μg/g all of the nymphs died without completing the molt. At 15 and particularly at 25 μg/g some nymphs never initiated a molt attempt (Category IV); in some cases such nymphs lived for over 60 days. These results parallel those obtained with *Locusta* (*15, 34*), *Oncopeltus* (*17*), and some other insects (*35-37*).

As *M. sanguinipes* nymphs will readily consume physiologically active doses of azadirachtin, we were able to compare the toxicity of this compound following oral, topical, and intrahemocoelic administration, and so evaluate the significance of the gut and integument as barriers to bioavailability (Figure 4). Azadirachtin is similar in toxicity when administered topically or by injection (ED_{50} = 4.5 and 3.2 μg/g insect fwt respectively; ED_{50} is the effective dose required to inhibit molting in 50% of the population over the length of the instar) suggesting that the integument does not limit bioavailability. Azadirachtin is however about three times less toxic when administered orally (LD_{50} = 11.3 μg/g insect fwt), so the gut limits its bioavailability to putative target sites within the insect. That this barrier is due to oxidative metabolism is suggested by the observation that oral toxicity is increased twofold by coadministration of 500 μg of piperonyl butoxide, a specific inhibitor of the mixed function oxidase system (Figure *4*). The results of these experiments suggest that doses of up to 8 μg/g might be excluded from reaching the hemolymph by oxidative metabolism, but at higher doses sufficient intact azadirachtin is absorbed from the gut to affect molting.

In addition to molt disruption, azadirachtin decreases the growth rate of *M. sanguinipes* nymphs, largely by decreasing the consumption of food (RCR) (Table III). This is of interest as our earlier experiments established that *M. sanguinipes* showed no antifeedant response to azadirachtin within a single feeding bout; the decreased RCR may then be due to direct toxic effects on the gut or neural inhibition of feeding rather than to peripheral

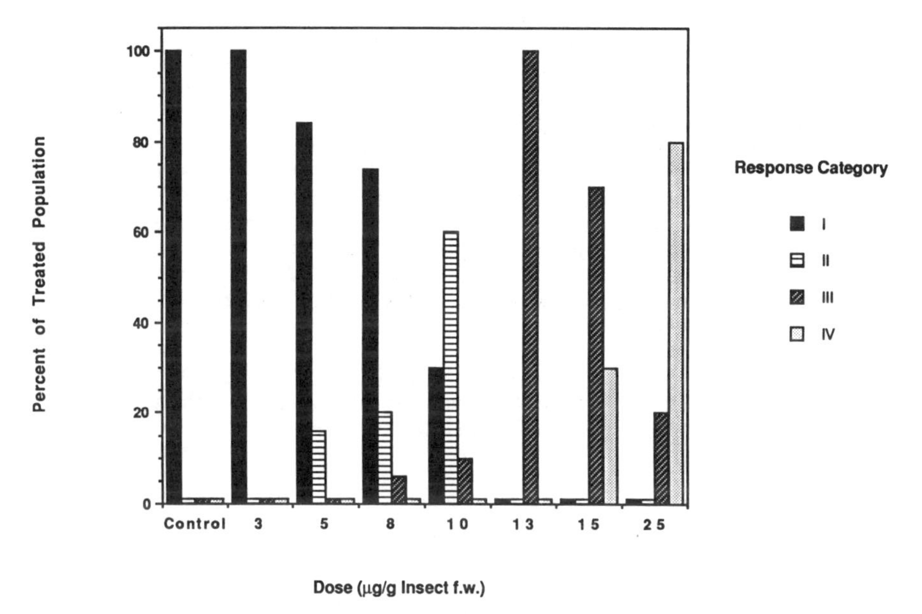

Figure 3. Effect of orally administered azadirachtin on molting of fifth instar *Melanoplus sanguinepes* nymphs. Response categories are no effect (I), crumpled wings or tarsi (II), died in a failed molt (III), and died without molting (IV).

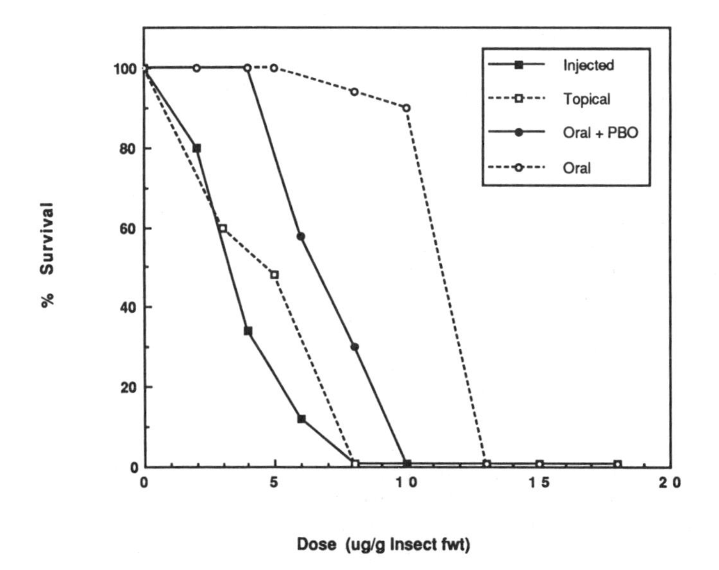

Figure 4. Comparison of azadirachtin effects on *M. sanguinipes* molting following application orally, orally with 500 ug piperonyl butoxide (PBO), topically, and by injection. In this figure category I and II respones are pooled to give survivorship curves.

chemosensory responses. As noted earlier in the case of *P. saucia*, the increased AD may be due to the increased tenure of the food bolus as it passes through the alimentary canal. Both the decreased RCR and increased AD are consistent with a decrease in the rate of gut peristalsis, as demonstrated by Mordue *et al.* (*31*).

This suggests that reports of the antifeedant activity of azadirachtin on various insects must be interpreted with caution if the assay used is conducted for a period of time longer than that of a single feeding bout. On the other hand electrophysiological experiments and choice tests do show that some insects have an obvious chemosensory based response to azadirachtin (*38, 39*). Azadirachtin also decreases the ECI and ECD in *M. sanguinipes*, indicating that the treated individuals are under metabolic stress prior to the initiation of molting.

Table III. Effect of azadirachtin on dietary utilization by fifth instar *Melanoplus sanguinipes*. Values given are the mean ± 1 standard deviation

Dose μg/g insect	RGR (mg/mg/d)	RCR (mg/mg/d)	AD (%)	ECI (%)	ECD (%)
Control	0.24±.04	0.83±.20	50.1±13.4	29.3± 9.1	70.4±42.6
10 μg/g	0.14±.03	0.58±.30	56.4±16.9	36.8±23.0	55.0±19.8
15 μg/g	0.13±.04	0.53±.09	60.2±09.9	26 2± 8.0	44.8±15.7

RGR is the Relative Growth Rate, RCR is the Relative Consumption Rate, AD is the Approximate Digestibility, ECI is the Efficiency of Conversion of Ingested diet, and ECD is the Efficiency of Conversion of Digested diet. All values were calculated according to Reese and Beck (1976).

Several studies have established that azadirachtin inhibits molting by interfering with the production of ecdysteroids prior to apolysis. However, the precise mechanism by which this occurs is unknown. Azadirachtin does not block the ability of prothoracic glands *in vitro* to produce ecdysone or respond to prothoracicotrophic hormone (*40, 41*), and does not compete with ecdysone for receptor binding sites (*42*). It has been suggested that azadirachtin's effect may involve targets within the gut, disrupting some process necessary to trigger the molting processes (*34*). We noted the similarity of toxicology of azadirachtin and the azasterols; both classes of compounds produce growth and molt inhibition, and chemosterilization at doses in the 1-10 ppm range (*43*). The azasterols are known to inhibit the Δ^{22} and $\Delta^{22,24}$ sterol reductase enzymes, and so block the conversion of sitosterol to cholesterol, resulting in the accumulation of the metabolic intermediate desmosterol (*43, 44*). Azasterol toxicity can be completely reversed by supplementing the insect's diet with cholesterol. We were unable to reverse azadirachtin toxicity by supplementing the diet with 1,000 ppm cholesterol, suggesting that the sterol reductases are not a target for azadirachtin.

The possibility that azadirachtin interferes with the transport of sterols through the hemolymph was also examined. Azadirachtin-treated and control fifth instar *M. sanguinipes* were fed ^{14}C-β-sitosterol, after which 10 μl hemolymph samples were taken at hourly intervals for 12 hours. Radiolabel appeared in the hemolymph of both treatment groups 4 hours after feeding. In the controls the amount of radioactivity detected peaked at 9-10 hours after feeding and subsequently declined rapidly, whereas in the azadirachtin treated group radioactivity increased more gradually and did not peak within the 12 hour duration of the experiment. As radiolabelled sterol appeared in the hemolymph in both groups azadirachtin does not appear to block the action of the carrier lipoproteins involved in sterol transport.

Comparative Toxicity Studies

We have assayed nine other limonoids for feeding and growth inhibition against *P. saucia*, and for molting inhibition against *Oncopeltus fasciatus*, a species particularly sensitive to azadirachtin (*45*). Cedrelone and anthothecol are two relatively simple limonoids with a wide distribution in the Meliaceae; they differ only in that anthothecol has an acetate sustitution at the C11 position of the C ring. Both compounds display similar growth inhibiting properties against *P. saucia* larvae at 0.5 μmol/g diet fwt (Table IV). However only cedrelone has molt inhibiting activity when topically applied to *O. fasciatus* nymphs (Figure 5). The symptoms produced are similar to those caused by azadirachtin, but the ED_{50} (50% molt inhibition) is about 20 μg/nymph compared to 0.02 μg/nymph for azadirachtin. Cedrelone also inhibits molting in the noctuid *Heliothis zea* (*3*). The substituent at the C11 position affects the biological activity, as anthothecol had no molting inhibitory effect at doses up to 50 μg/nymph.

Table IV. Effect of limonoids (at 0.5 μol/g diet fwt) on neonate *P. saucia* growth and diet choice. Asterisks indicate values which differ significantly from the control (Duncan's Multiple Range test, $p \leq .05$ for growth data; X^2, $p \leq .05$ for diet choice)

Compound	Growth (% of Control) (mean ± S.D.)	Percentage of larvae on Control (C) or Treated (T) diet	
		C	T
Cedrelone	10.8 ± 3.2*	57	43
Anthothecol	12.9 ± 7.5*	57	43
Harrisonin	89.2 ± 22.6	80	20*
Obacunone	110.8 ± 11.8	61	39
Nomilin	118.3 ± 18.3	63	37
Gedunin	96.8 ± 23.7	43	57
Pedonin	133.3 ± 8.6*	63	37
Entandrophragmin	96.8 ± 18.3	66	34
Bussein	64.5 ± 19.5*	67	33

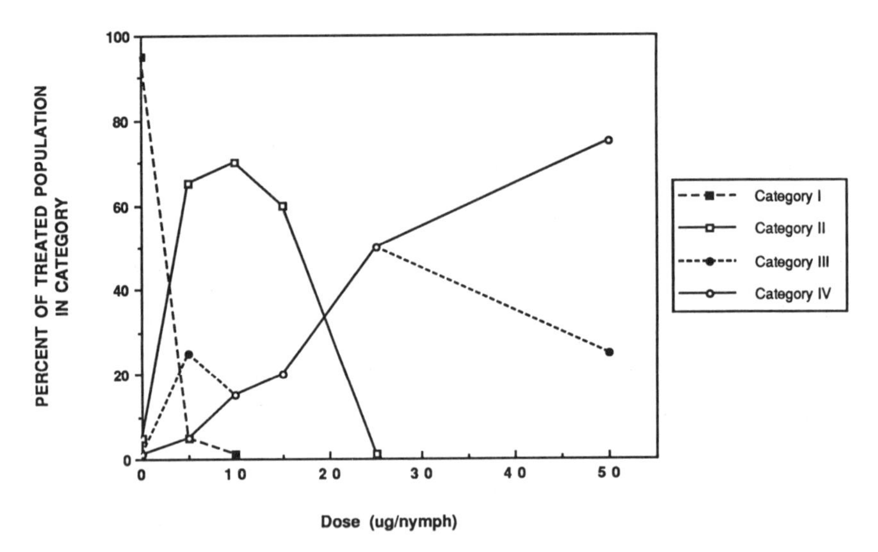

Figure 5. Effect of topically applied cedrelone on the molting of *Oncopeltus fasciatus* nymphs. Response categories are no effect (I), crumpled wings or tarsi (II), died in a failed molt (III), and died without molting (IV).

Many limonoids of the Rutaceae and Simaroubaceae, and some from the Meliaceae, possess a characteristic D ring structure involving formation of a lactone and epoxidation at the 14, 15 position (D-seco limonoids) (*46*). We compared the activities of four such compounds: harrisonin, from *Harrisonia abyssica*, obacunone and nomilin, characteristic Rutaceous limonoids, and gedunin, a compound which occurs in many species of Meliaceae. None of the compounds inhibited *O. fasciatus* molting or reproduction at doses up to 50 μg/nymph. Harrisonin showed antifeedant activity against neonate *P. saucia* at 0.5 μmol/g diet fwt, but the larvae were apparently able to compensate for this subsequently as this compound did not inhibit growth at the end of a seven-day trial. The other three compounds showed no activity against *P. saucia*. These results contrast with the reported activity of citrus limonoids against other species of Noctuidae (*4*), but previous studies with harrisonin and other limonoids indicate that the response may vary widely between even closely related species (*3, 8*). A related compound, pedonin (*9*), with an opened D ring and a rearranged A ring, was stimulatory to *P. saucia* neonates at 0.5 μmol/g.

Many Meliaceous limonoids, including azadirachtin, have undergone extensive oxidation and rearrangement; indeed this is a characteristic evolutionary pattern within the family (*47*). We tested two such compounds, entandrophragmin and bussein. Neither compound inhibited *Oncopeltus* molting at 50 μg/nymph. Both showed weak antifeedant activity against *P. saucia* neonates, but only bussein at 0.5 μmol/g diet fwt resulted in significant growth inhibition.

These results do not point to common structural features necessary for insecticidal or antifeedant activity. The differential activity of cedrelone and anthothecol may be due to the slight difference in polarity produced by the extra acetate function, but cedrelone and azadirachtin, the only two compounds to produce molt inhibition in our assays, differ widely in polarity. Our results compare with the earlier surveys of Kubo and Klocke (*3*) and Arnason *et al.* (*7*) who also found cedrelone and anthothecol to have insect growth inhibiting activity. The exceptional activity of azadirachtin in relation to the other limonoids is emphasized in all of these studies.

Plant Extract Screening

We have assayed extracts of a variety of species of Meliaceae for feeding deterrency and growth inhibition against *P. saucia* (Table V) (*48*). Methanolic extracts of foliage were incorporated into an artificial diet to achieve concentrations of 25, 50, 75, and 100% of natural leaf concentration, and larvae were allowed to feed for seven days before weight gain and survival were assessed. In general, species of the Melioideae produced more toxic extracts than species of the Swietenioideae, which is somewhat surprising as the latter is considered to be the more evolutionarily advanced subfamily based on morphology (*49*). The production of insecticidal extracts does not follow well-defined taxonomic lines, and is more likely to be related to ecological conditions idiosyncratic to particular species.. Extracts of some species, including *Aglaia odorata* and *Turreae holstii*, are almost as active as those of *Azadirachta indica*. The phytochemistry of these species has been investigated and will be reported elsewhere.

Table V. Bioactivity of meliaceous leaf extracts on neonate *P. saucia*. Values given are the concentration (as % of natural leaf concentration) of the total MeOH extract administered in artificial diet required to reduce growth (EC_{50}) or survivorship (LC_{50}) by 50% relative to the control, over a seven day assay

	EC_{50}	LC_{50}
Family Meliaceae		
Subfamily Melioideae		
Tribe 1. Turreeae		
Turreae holstii	2.5	63
Tribe 2. Melieae		
Melia azediracht	2.5	5.0
Melia toosenden	2.5	5.0
Azadirachta indica	2.5	5.0
Tribe 4. Trichilieae		
Trichilia hirta	12.5	>100
Lepidotrichilia volkensii	12.5	>100
Ekebergia capensis	43.0	>100
Cipadessa baccifera	44.0	>100
Tribe 5. Aglaieae		
Aglaia odorata	1.7	15.0
Aglaia odoratissima	11.0	75.0
Aglaia argentia	27.0	>100
Aphanamixus grandifolia	>100	>100
Tribe 6. Guareeae		
Guarea glabra	62.0	>100
Subfamily Swietenioideae		
Tribe 1. Cedreleae		
Cedrela odorata	67.0	>100
Cedrela serrata	27.0	>100
Toona ciliata australis	12.0	>100
Tribe 2. Swietenieae		
Khaya senegalensis	35.0	75.0
Chuckrassia tabularis	12.5	>100
Swietenia humilis	23.0	>100
Swietenia mahogani	17.0	>100
Swietenia candollei	20.0	80.0

Acknowledgments

Dr. A. Hassanali, ICIPE, Nairobi, Dr. I. Kubo, University of California, and Dr. J.T. Arnason, University of Ottawa, provided samples of limonoids for bioassay. Dr. K.R. Downum, Florida International University, and Mr. T. Flynn, Pacific Tropical Gardens, Hawaii, provided leaf samples. This work was supported by NSERC operating grants to M.B. Isman (A2729) and G.H.N. Towers (A2143).

Literature Cited

1. Balandrin, M.F., J.A. Klocke, E.S. Wurtele, and W.H. Bollinger 1985. Science 228:1154-1160.
2. Hedin, P.A. 1982. J. Agric. Food Chem. 30:201-215.
3. Kubo, I., and J. A. Klocke 1986. In Natural Resistance of Plants to Pests. (Ed. by M.B. Green and P.A. Hedin), pp. 206-219, American Chemical Society Symposium Series 296, Washington, D.C.
4. Klocke, J.A., and I. Kubo 1982. Entomol. exp. appl. 32:299-301.
5. Morgan, E.D. 1982. In Natural Pesticides from the Neem Tree (*Azadirachta indica* A. Juss). Proc. 1st Int. Neem Conf. (Rottach-Egern, 1980), (Ed. by H. Schmutterer, K.R.S. Ascher, and H. Rembold), pp. 43-52. German Agency for Technical Cooperation, Eschborn, Germany.
6. Butterworth, J.H., and E.D. Morgan 1971. J. Insect Physiol. 17:969-977.
7. Arnason, J.T., B.J.R. Philogene, N. Donskov, and I. Kubo 1987. Entomol. exp. appl. 43:221-226.
8. Hassanali, A., M.D. Bently, E.N. Ole Sitayo, P.E.W. Njoroge, and M. Yatagai 1986. Insect Sci. Applic. 7:495-499.
9. Hassanali, A., M.D. Bentley, A.M.Z. Slawin, D.J. Williams, R.N. Shephard, and A.W. Chapya 1987. Phytochemistry 26:573-575.
10. Lidert, Z., D.A. Taylor, and M. Thirugnanam 1985. J. Nat. Prod. 48:843-845.
11. Koul, O. 1983. Z. Ang. Ent. 95:166-171.
12. Kraus, W., W. Grimminger, and G. Sawitzki 1978. Agnew. Chem. Int. Ed. Engl. 17:452-453.
13. Nakatani, M., M. Okamoto, T. Iwashita, K. Mizukawa, H. Naoki, and T. Hase 1984. Heterocycles 22:2335-2340.
14. Sieber, K.-P., and H. Rembold 1983. J. Insect Physiol. 29:523-527.
15. Mordue, A.J., K.A. Evans, and M. Charlett 1986. Comp. Biochem. Physiol. 85C:297-301.
16. Schluter, U., H.J. Bidmon, and S. Grewe 1985. J. Insect Physiol. 31:773-777.
17. Redfern, R.E., T.J. Kelly, A.B. Borkovec, and D.K. Hayes 1982. Pest. Biochem. Physiol. 18:351-356.
18. Dorn, A., J.M. Rademacher, and E. Sehn 1986. J. Insect Physiol. 32:231-238.
19. Rembold, H., H. Forster, Ch. Czoppelt, P.J. Rao, and K.P. Sieber 1984. In Natural Pesticides from the Neem Tree (*Azadirachta indica* A. Juss) and Other Tropical Plants. Proc. 2nd Int. Neem Conf. (Rauischholzhausen, 1983), (Ed. by H. Schmutterer, and K.R.S. Ascher), pp. 163-179. German Agency for Technical Cooperation, Eschborn, Germany.

20. Koul, O., K. Amanai, and T. Ohtaki 1987. J. Insect Physiol. 33:103-108.
21. Schmutterer, H. 1988. J. Insect Physiol. 34:713-719.
22. Jacobson, M. 1986. In Natural Resistance of Plants to Pests. (Ed. by M.B. Green and P.A. Hedin), pp. 220-232. American Chemical Society, Symposium Series Volume 296, Washington, D.C.
23. Okpanyi, S.N. and G.C. Ezeukwu 1981. Planta Med. 41:34-39.
24. Arnason, J.T. 1988 personal communication.
25. Saxena, R.C., H.D. Justo, and P.B. Epino 1983. In Natural Pesticides from the Neem Tree (*Azadirachta indica* A. Juss) and Other Tropical Plants. Proc. 2nd Int. Neem Conf. (Rauischholzhausen, 1983), (Ed. by H. Schmutterer, and K.R.S. Ascher), pp. 163-179. German Agency for Technical Cooperation, Eschborn, Germany.
26. Champagne, D.E., M.B. Isman, and G.H.N. Towers, in prep.
27. Reese, J.C. and S.D. Beck 1976. Ann. Ent. Soc. Amer. 69:59-67.
28. Arnason, J.T., B.J.R. Philogene, N. Donskov, C. McDougall, G. Fortier, P. Morand, D. Gardner, J. Lambert, C. Morris, and C. Nozzolillo 1985. Entomol. exp. appl. 38:29-34.
29. Barnby, M.A. and J.A. Klocke 1987. J. Insect Physiol. 33:69-75.
30. Fagoonee, I. 1984. In Natural Pesticides from the Neem Tree (*Azadirachta indica* A. Juss) and Other Tropical Plants. Proc. 2nd Int. Neem Conf. (Rauischholzhausen, 1983), (Ed. by H. Schmutterer, and K.R.S. Ascher), pp. 211-224. German Agency for Technical Cooperation, Eschborn, Germany.
31. Mordue, A.J., P.K. Cottee, and K.A. Evans 1985. Physiol. Entomol. 10:431-437.
32. Mulkern, and Mongolkiti 1975. Acrida 4:95
33. Champagne, D.E., M.B. Isman, and G.H.N. Towers, in prep.
34. Mordue, A.J. and K.A. Evans 1987. In Insects-Plants, pp 43-48, Labeyrie, V., Fabres, G., Lachaise, D., Eds. Junk, Dordrecht, Netherlands.
35. Koul, O. 1984. Zeitschrift Ange. Entomol. 98:221-223.
36. Ladd, T.L.Jr., Warthen, J.D.Jr., and M.G. Klein 1984. J. Econ. Entomol. 77:903-905.
37. Gaaboub, I.A., and D.K. Hayes 1984. Environ. Entomol. 13:1639-1643.
38. Schoonhoven, L.M. 1982. In Pesticides from the Neem Tree (*Azadirachta indica* A. Juss). Proc. 1st Int. Neem Conf. (Rottach-Ergen, 1980), (Ed. by H. Schmutterer, K.R.S. Ascher, and H. Rembold), pp. 105-108. German Agency for Technical Cooperation, Eschborn, Germany.
39. Simmonds, M.S.J., and W.M. Blaney 1984. In Natural Pesticides from the Neem Tree (*Azadirachta indica* A. Juss) and Other Tropical Plants. Proc. 2nd Int. Neem Conf. (Rauischholzhausen, 1983), (Ed. by H. Schmutterer, and K.R.S. Ascher), pp. 163-179. German Agency for Technical Cooperation, Eschborn, Germany.
40. Koul, O., K. Amanai, and T. Ohtaki 1987. J. Insect Physiol. 33:103-108.
41. Pener, M.P., D.B. Roundtree, S.T. Bishoff, and L.I. Gilbert, 1988. In Endocrinological Frontiers in Physiological Insect Ecology. (Ed. by F. Senhal, A. Zabza and D.L. Denlinger), pp. 41-54. Wroclaw Technical University Press, Wroclaw, Poland.
43. Svoboda, J.A. and W.E. Robbins 1971. Lipids 6:113-119.
44. Svoboda, J.A. and W.E. Robbins 1967. Science 156:1637-1638.
45. Champagne, D.E., M.B. Isman, and G.H.N. Towers, in prep.

46. Dreyer, D.L. 1983. In Chemistry and Chemical Taxonomy of the Rutales, P.G. Waterman and M.F. Grundon, Eds. pg. 215-246. Academic Press, London.
47. Das, M.F., G.F.D. Silva, O.R. Gottlieb, and D.L. Dreyer 1984. Biochem. System. Ecol. 12:299-310.
48. Champagne, D.E., M.B. Isman, and G.H.N. Towers, in prep.
49. Pennington, T.D., and B.T. Styles 1975. Blumea 22:419-540.

RECEIVED November 18, 1988

Chapter 9

Insecticides from Neem

R. C. Saxena

Entomology Department, International Rice Research Institute, P.O. Box 933, Manila, Philippines

Derivatives of neem (Azadirachta indica A. Juss) have traditionally been used by farmers in Asia and Africa to ward off insect pests of household, agricultural, and medical importance. Unlike ordinary insecticides based on single active ingredients, neem derivatives comprise a complex array of novel compounds which have diverse behavioral and physiological effects on insects. Repellency, feeding and oviposition deterrence, growth and reproduction inhibition, and other effects have been attributed to neem compounds - azadirachtin, salannin, meliantriol, etc. that occur mainly in the seed. However, the complexity of chemical structure of these compounds precludes their synthesis on a practical scale. Therefore, the use of simple formulations of neem derivatives such as leaf or kernel powder or extracts needs to be popularized. Their being safe to non-target organisms, including humans, make them ideal insecticides. Several azadirachtin-rich formulations have already been commercialized for use on nonfood and food crops.

Insecticides are needed to control at least half of the insect problems affecting agriculture and public health (1). Crop losses would soar and food prices will escalate if insecticides were to be banned (2). Crop losses would be of even higher magnitude in developing countries where alternative crop protection measures are still being strengthened (3).

Since the advent of DDT, most insecticides that have been developed are synthetic, nonselective, and

0097–6156/89/0387–0110$07.50/0

toxic chemicals. Although they have effectively controlled some pest species, their extensive and sometimes indiscriminate use has led to serious social and environmental repercussions. The poisoning of livestock, fish, wildlife, and other beneficial organisms has been linked with increased insecticide use. There has also been a disturbing increase in human poisoning, particularly in developing countries where safe handling and application of insecticides is not always feasible due to several socioeconomic factors. Pest resurgence, associated with insecticidal destruction of natural enemies, and the widespread development of insecticide resistance in pests have warranted higher doses or more powerful insecticides, ending in an insecticide treadmill. That is not only uneconomical, it also exacerbates the problem. Clearly, new insecticides will have to meet entirely different standards. They must be pest-specific, nontoxic to humans and beneficial organisms, biodegradable, less prone to insect resistance and resurgence, and less expensive.

In the last two decades, interest has increased in the potential and possible use of new bioactive products and natural insecticides, which are less likely to cause ecological damage. These alternative insect control agents include botanical insecticides, microbial insecticides, insect metabolic disruptors and inhibitors, pheromones, and other modifiers of insect behavior. Insect growth regulators (IGRs), such as juvenile hormones (JH) and JH-analogs, were heralded as "third-generation pesticides" (4), while JH-antagonists, such as precocenes, were considered as "fourth-generation insecticides" (5). However, the practicality of pheromones and IGRs has been limited because of their high costs and apparent selectivity of certain species. The control potential of other bioactive candidates, such as insect neuroproteins and peptides (6,7) and antagonists of insect intracellular symbiotes (8) is being investigated.

Plants virtually are the richest source of bioactive organic chemicals on earth. Although only about 10,000 secondary plant metabolites have been chemically identified, the total number of plant chemicals may exceed 400,000 (9). They are a vast cornucopia of defense chemicals, comprising repellents, feeding and oviposition deterrents, growth inhibitors, sterilants, toxicants, etc. Many of the oldest and most common insecticides, such as nicotine, pyrethrins, and rotenone, were derived from plants. The chemical or insecticide approach had its beginnings in the use of botanical materials. This paper reviews the potential of neem, *Azadirachta indica* A. Juss (Family Meliaceae), as a source of natural insecticides.

Historical Perspective

The neem tree or margosa is indigenous to India (10). It is now widespread in many Asian and African countries and is also grown in Australia, Fiji, Papua New Guinea, the Philippines, Mauritius, several countries in Central and South America, the Caribbean, Puerto Rico, and the Virgin Islands. Its uses have been well known in India and are mentioned in the earliest Sanskrit medical writings (11). Almost every part of the tree is bitter but the seed kernel is most bitter. Centuries before synthetic insecticides became available, farmers in India protected crops with natural repellents found in neem fruits and leaves (12). Even now dried neem leaves are mixed with grain in Indian households to provide protection against storage pests. Likewise, the use of neem leaves for protecting woollen clothes from insect damage is not uncommon. Neem cake, a bulky residue of neem seeds after oil extraction, was applied to ricefields to protect the crop against insect pests as early as 1932 in India (13). Neem application rendered the paddy bitter, thus discouraging pest attack. The cake was also reported to serve as a manure and as a preventive measure against white ants.

Pradhan et al. (14) discovered the antifeedant activity of neem seeds against locusts. Lavie et al. (15) identified meliantriol as a locust phago-repellent from neem. A year later, Butterworth and Morgan (16) isolated azadirachtin, the most potent and chemically the most important antifeedant principal from neem seeds. But even after such a long history of neem's pest control potential, its use declined. This paradox can be attributed to the researchers' pre-dilection for insecticides having a distinctive toxicity as epitomized by DDT and a whole array of broad spectrum, synthetic insecticides. Only recently have the potentials of behavioral and physiological aberrations been recognized. They are highly desirable in integrated pest management programs as they minimize the risk of exposing the pests' natural enemies to poisoned food or starvation. Within the last decade, neem has come under close scrutiny of scientists around the world as the most promising source of natural insecticides (17-20). Since 1980, eight national and international conferences have been held in West Germany, India, the Philippines, Kenya, and Australia to review neem's pest control potential.

Chemistry of Neem Compounds

The chemistry of neem compounds has been reviewed by Jacobson (21-23), Kraus et al. (24,25), and Morgan

(26). The active principals or the "bitters" have been identified as limonoids, a group of stereochemically homogeneous tetranortriterpenoids. The most important active principal is azadirachtin, although more than 25 different compounds have been isolated including beta-sitosterol, fatty acids, and flavonoids. At least 9 compounds possess insect growth regulating activity. Besides azadirachtin, meliantriol (15) and salannin (27) are also active feeding deterrents. A new limonoid, deacetyl-azadirachtinol, isolated from fresh fruits, was as effective as azadirachtin in assays against the tobacco budworm *Heliothis virescens* (F.) (23). Also, another antifeedant and growth inhibitor closely related to azadirachtin (designated as vepaol) and two other apparently new compounds (designated as isovepaol and nimibidin) have been isolated from neem seed (28). Recently, a new limonoid insect growth inhibitor (7-deacetyl-17-hydroxy-azadiradione) has also been isolated from neem (29).

Meliantriol and salannin have been obtained in discrete crystalline form and structurally defined, but there has been much inconsistency about the chemical structure of azadirachtin since till recently, it could be obtained only as a white amorphous powder that melts at 154-158°C. Broughton et al. (30), through X-ray crystallography of a detigloylated (tigloyl=2-methyl-crotonyl) dihydro derivative of azadirachtin, have now conclusively proposed the following molecular structure of azadirachtin (Fig.1):

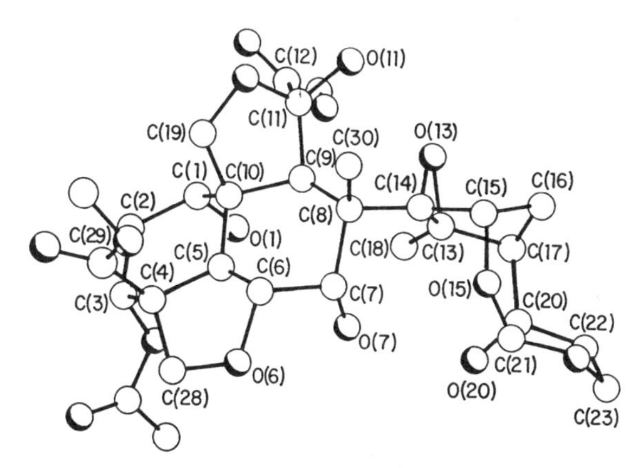

Fig. 1. Molecular structure of azadirachtin (30)

Neem Products with Insect Control Potential

The complexity of neem compounds precludes their synthetic production in the near future. Consequently, most neem derivatives used or evaluated for insect control include dried leaves, seed, seed kernel, oil, cake, aqueous or organic solvent extracts of seed kernel, standardized "Neemrich" extracts, partially purified fractions, azadirachtin-rich formulations, etc.

The thick, viscous, crude neem oil has a strong garlic odor due to the presence of a large number of sulfur moieties. India produces approximately 83,000 tons of neem oil and about 332,000 tons of neem cake annually (17, 31, 32). The major outlet for neem oil in India is the soap industry where it is used in the manufacture of low-grade soaps and detergents. It is also used as a lubricant and as a paint thinner.

Neem cake is in great demand for pest control use, especially against plant nematodes in cardamom plantations in Kerala, where approximately 3,000 metric tons of neem cake is being sold annually, mostly by pesticide dealers (Ahmed, S., East West Center, Honolulu, USA, personal communication, 1988.). Neem cake is also valuable as a nitrification inhibitor and is being used for coating or blending with urea fertilizer (33).

A company in India is marketing two neem-based formulations -- Repelin and Wellgro -- for spraying in tobacco-growing areas of Andhra Pradesh to check damage by cutworms, other insect pests, and tobacco mosaic virus in tobacco nurseries (Subramaniam, T. S., I.T.C., Ltd., Rajahmundry, India, personal communication, 1987.) Use of Repelin is also becoming popular in cotton-growing regions in Andhra Pradesh (Ahmed, S., East West Center, Honolulu, USA, personal communication, 1988), where cotton damage by the cotton bollworm *Heliothis armigera* Hübner was estimated at US$40 million (Jayaraj, S., Tamil Nadu Agricultural University, Coimbatore, India, personal communication, 1988), and where repeated applications of synthetic pyrethroids led to the resurgence of whiteflies. Another company is marketing two neem oil-based formulations -- Nimbosol and Biosol -- for the control of whiteflies and lepidopterous pests (Kather, M. A., A. V. Thomas & Co., Ltd., Madras, India, personal communication, 1988.) Registration has also been granted in the USA for the use of an azadirachtin-rich formulation Margosan-O for use on nonfood crops and ornamentals (34). Use on food crops may also be allowed shortly as much data on the safety of neem derivatives to man and beneficial organisms have been generated. An Indian company has come up with an azadirachtin-rich granular formulation Neemark (Khandal, V. S., West Coast Herbochem, Bombay, India, personal communication, 1988.) Neemark is recommended for use (at a rate of 18.75 kg/ha) on cotton, paddy, tobacco, groundnut, sugarcane, chili, eggplant, other common vegetables, horticultural crops, legumes, and food grains.

Researchers at the National Chemical Laboratory in India have advocated the development and use of neem extracts enriched in particulate principals or properties as opposed to single, active ingredient for insect control (35). These extracts can be produced on a cottage industry level or by small-scale extraction plants at the village level. A standardized extract "Neemrich I" has been produced, probably the first of its kind to be registered for use. The extraction process requires only a few steps. Two additional extracts "Neemrich II" and "Neemrich III" are also being developed. Contrary to the conventional practice of providing a precise chemical description of the active ingredient, the Neemrich extracts would have to be judged with respect to their biological activity. This necessitates a complete revision and reconciliation of registration concepts and policies. Although it is possible to describe such mixtures chemically, the cost will be prohibitive.

Recently, a simple procedure has been developed at the International Rice Research Institute to extract "bitters" (limonoids) from neem seed kernel as a crystalline brownish powder (36). The powder, unlike neem oil, is water soluble, relatively photostable, and nonphytotoxic.

Range of Insect Species Affected

In spite of neem's high selectivity, neem derivatives affect a wide range of insect pests. Jacobson (22) listed 123 insect species belonging to orders Coleoptera, Diptera, Heteroptera, Homoptera, Hymenoptera, Lepidoptera, and Orthoptera, in addition to 3 mite and 5 nematode species. Since then, neem derivatives have been found to affect 75 additional insect species belonging to different orders, and an ostracod (Table I).

Biological Effects and Modes of Action of Neem Derivatives

Neem derivatives have diverse behavioral and physiological effects on insects ranging from repellency to feeding deterrence, growth disruption, sterilizing effects, mating disruption, oviposition inhibition, etc. At times, elucidation of the precise mode of action is difficult because of the complex array and concerted or synergistic effects of neem compounds present in the leaves, bark, seed, oil, cake, or extracts. However, the effects have been categorized for better understanding.

Table I. Additional species of arthropods affected by neem derivatives as cited in selected references[a]

Scientific name	Common name	Reference
COLEOPTERA		
Callosobruchus analis (Fab.)	pulse beetle	37
Dicladispa armigera Olivier	rice hispa	38
Epitrix fuscula Crotch	eggplant flea beetle	39
Madurasia obscurella Jacoby	galerucid beetle	40
Henosepilachna vigintioctopunctata (Fab.)	coccinellid beetle	41
Myllocerus sp.	cotton grey weevil	28
Ootheca bennigseni Weise	foliar beetle	42
Schematiza cardiae Barber	beetle	43
DIPTERA		
Aedes togoi (Theobald)	mosquito	44
Calliphora vicina R.-D.	blue blowfly	45
Culex quinquefasciatus Say	southern house mosquito	44
Dacus cucurbitae Coquillet	melon fly	46
Melanagromyza obtusa (Mall.)	bean pod fly	47
Musca domestica L.	house fly	48
Ophiomya phaseoli (Tryon)	bean fly	49
Orseolia oryzae (Wood-Mason)	gall midge	50
Phormia regina (Meigen)	black blowfly	51
Phormia terraenovae R.-D.	blowfy	51
HETEROPTERA		
Antestiopsis orbitalis bechuana (Kirk.)	East African coffee bug	52
Calocoris angustatus Leth.	sorghum earhead bug	28
Leptocorisa oratorius F.	rice bug	53
Scotinophora coarctata F.	Palawan black bug	54
HOMOPTERA		
Amrasca devastans (Distant)	cotton leafhopper	55
Aphis citricola van der Goot	citrus aphid	56
Aphis gossypii Glover	cotton aphid	57
Bemisia tabaci (Gennadius)	sweetpotato whitefly	58
Diaphorina citri Kuway	citrus psyllid	59
Empoasca fascialis (Jac.)	cotton jassid	60
Empoasca lybica de Berg.	cotton jassid	57
Jacobiella facialis (Jac.)	eggplant leafhopper	60
Lipaphis erysimi (Kalt.)	mustard aphid	61
Melanaphis sacchari (Zehnt.)	sugarcane aphid	62
Myzus persicae (Sulzer)	green peach aphid	63
Rhopalosiphum maidis (Fitch)	corn leaf aphid	62
Rhopalosiphum nymphaeae (L.)	water lily aphid	64
Toxoptera aurantii (Boyer)	black citrus aphid	59
HYMENOPTERA		
Athalia lugens proxima (Klug)	mustard sawfly	65

Table I. Continued

Scientific name	Common name	Reference
Fenusa pusilla (Lepeletier)	birch leafminer	66
Formica polyctena Foerster	wood ant	67
LEPIDOPTERA		
Achaea janata L.	castor semilooper	68
Amsacta moorei Butler	redhairy caterpillar	69
Boarmia selenaria (Schiff.)	giant looper	70
Chilo partellus (Swinhoe)	spotted stem borer	28
Corcyra cephalonica (Stain.)	rice moth	71
Crocidolomia binotalis Zel.	cabbagehead caterpillar	72
Earias fabia Stoll	spotted bollworm	73
Euchrysops cnejus (F.)	blue butterfly	74
Heliothis armigera Hübner	cotton bollworm	75
Hellula undalis (F.)	cabbage webworm	60
Maliarpha separatella Rag.	white borer	76
Maruca testulalis (Geyer)	pod borer	77
Ostrinia furnacalis (Guenée)	Asiatic corn borer	77
Ostrinia nubilalis (Hubner)	European corn borer	78
Phthorimaea operculella (Zell.)	potato tuber moth	35
Phycita melongenae Aina	leaf roller	60
Phyllocnistis citrella Stain.	citrus leafminer	79
Pieris brassicae (L.)	cabbage butterfly	80
Pieris rapae (L.)	imported cabbageworm	81
Scirpophaga incertulas (Wlk.)	yellow stem borer	50
Scrobipalpa ergasima Mayr	tobacco stem borer	60
Selepa docilis Butl.	noctuid	60
Sesamia calamistis Hmps.	stem borer	60
Sesamia nonagrioides (Lef.)	stem borer	60
Spilosoma obliqua (Wlk.)	jute hairy caterpillar	74
Spodoptera eridania (Cram.)	southern armyworm	82
Spodoptera mauritia acronynctoides (Boisd.)	rice armyworm	31
Syllepte derogata (F.)	okra leafroller	83
ORTHOPTERA		
Blatta orientalis L.	oriental cockroach	84
Blattella germanica (L.)	German cockroach	84
Byrsotria fumigata G.-M.	cockroach	84
Gromphadorhina portentosa (Sch.)	cockroach	84
Periplanata americana (L.)	American cockroach	84
Supella longipalpa (F.)	brownbanded cockroach	84
Zonocerus variegatus L.	variegated grasshopper	85
THYSANOPTERA		
Stenchaetothrips biformis (Bag.)	rice thrips	86
OSTRACODA (Class Crustacea)		
Heterocypris luzonensis Neale	ostracod	87

[a]Since Jacobson (22). Although there are several reports on any given species, only one reference is selected for brevity.

Repellent and Antifeedant Effects. Pradhan et al. (14) first demonstrated that a 0.001% aqueous suspension of crushed neem kernel sprayed on cabbage plant totally stopped feeding by the desert locust Schistocerca gregaria Forsk. on treated foliage. Neem treatment also deterred the migratory locust Locusta migratoria (L.) but at a concentration of 0.01% suspension. Crops sprayed with a 0.1% neem seed suspension at 300-600 liters/ha escaped damage from an invasion of a locust swarm in Delhi in 1962; unsprayed crops were devastated. Ketkar (17) listed 95 publications on insect repellent and antifeedant effects of neem derivatives. Later studies on additional pest species have confirmed the feeding inhibition caused by neem derivatives (22). However, in many of these studies, repellency was not clearly differentiated from feeding deterrence. According to Dethier et al. (88), the term "repellent" should be used only when there is an oriented locomotion away from the source of stimulus.

The repellent and antifeedant effects of neem derivatives on rice insect pests have been studied indepth. Rice plants sprayed with neem oil using an ultra-low volume (ULV) spray applicator were unattractive to the brown planthopper Nilaparvata lugens (Stål), the whitebacked planthopper Sogatella furcifera (Horváth), the leaffolder Cnaphalocrocis medinalis (Guenée), the ear-cutting caterpillar Mythimna separata (Wlk.), and the rice armyworm Spodoptera mauritia acronynctoides (Bois.) (31,32,53, 89,90). Only 36% of N. lugens females alighted on plants sprayed with 3% neem oil. Insect arrival decreased as the concentration of oil increased. The oil repelled N. lugens and S. furcifera adults even without contact. Likewise, few first-instar larvae of foliage feeders settled on leaf cuts treated with neem oil. Neem oil repelled the green leafhopper Nephotettix virescens (Distant) to a lesser degree than it did the planthoppers (53). The settling response of N. lugens females on rice plants grown in soil incorporated with urea, sulfur-coated urea, or neem cake-urea mixture did not differ (91).

Feeding by foliage feeders as well as sucking insects was deterred on neem-treated rice plants. In a 60-min observation, even starved N. lugens females avoided alighting on seedlings treated with 50% or 100% neem oil for 15 to 20 min. They were restless upon alighting and spent more than 30 min searching for feeding sites (89). The feeding duration decreased by 0.93 min/h for every 1% increase in oil concentration, while the search for feeding sites was intensified. Food intake decreased significantly on plants sprayed with 12% neem oil, but the insect fed normally on plants sprayed with a vegetable oil. Feeding by S. furcifera and N. virescens was also reduced on neem-treated plants (53,92).

Phloem feeding by N. virescens, monitored with an electronic device, on rice plants sprayed with neem oil was significantly less than on acetone-treated plants (93). Phloem feeding was erratic on neem-treated plants while, probing, salivation, and xylem feeding increased. Phloem feeding also decreased on plants kept in an arena permeated with neem oil odor (94).

Gill & Lewis (95) demonstrated neem's systemic antifeedant action. Young bean plants systemically treated with an aqueous neem kernel suspension, ethanolic extract, or pure azadirachtin at 1000, 100, and 10 ppm, respectively, were hardly damaged by Schistocerca adults. Systemic antifeedant action was also evidenced by low food intake by N. lugens females on rice plants grown in soil incorporated with neem cake (91). N. virescens ingested less from the phloem and more from the xylem of neem cake-treated plants (96). The increased xylem feeding probably served to offset desiccation resulting from repeated probing, profuse salivation, and restlessness.

Contact with azadirachtin disrupts food intake and increases the locomotory activity of insects (97). The antifeedants may either stimulate the insects' specialized deterrent receptors or interact with receptors sensitive to phagostimulants (98). The resulting distortion of normal receptor activity may inhibit food intake.

Growth Inhibition. The growth inhibitory effects of neem derivatives are much more profound than repellency or phagodeterrency. Leuschner (52) reported that topical application of a crude methanolic extract of neem leaves at 50 ug to 4th- or 5th-instar nymphs of the East African coffee bug Antestiopsis orbitalis bechuana (Kirk.) caused morphogenetic defects similar to those produced by natural or synthetic hormonally active substances. The treatment of 4th-instar nymphs caused abnormal wing cases, scutelli, and hemelytra in the succeeding instar. Crude methanolic extract of neem leaves or seed, neem oil, and purified fractions of seed kernel inhibited the growth of the Mexican bean beetle Epilachna varivestis Mulsant, the Colorado potato beetle Leptinotarsa decemlineata (Say), and the diamondback moth Plutella xylostella (L.) (99,100).

Ruscoe (101) reported that growth of P. xylostella, the cabbage butterfly Pieris brassicae (L.), H. virescens, and of the cotton stainer Dysdercus fasciatus Sign. was affected if larvae or nymphs contacted azadirachtin-treated substrates or fed on treated food. P. brassicae pupae were small and deformed. A neem fraction or azadirachtin caused metamorphic deviations in larvae of E. varivestis, the Mediterranean flour moth Ephestia kühniella Zell., and the honeybee Apis mellifera L., independent of feeding inhibition (102,103). E. varivestis larvae fed on neem-treated bean leaves became discolored, developed black legs, and could not pupate (104). The dark appearance of the larvae was due to the accumulation of melanin (105). Larvae of the fall armyworm Spodoptera frugiperda (J. E. Smith), the corn earworm Heliothis zea (Boddie), and the pink bollworm Pectinophora gossypiella (Saunders) fed on azadirachtin-treated artificial diet also, failed to molt (106).

The growth inhibitory effects of neem derivatives are now reported for numerous insect species. Thus nymphs of rice leafhoppers and planthoppers and the rice bug caged on neem oil- or neem extract-treated rice plants and lepidopterous larvae fed on treated leaves suffered from ecdysial failures and other developmental defects (53,89,90, 107). N. lugens biotypes 1, 2, and 3 were equally sensitive to neem treatment (89). The leafhoppers and planthoppers were also sensitive to systemically applied azadirachtin or extracts of neem seed kernel (36,108). The systemic effect persisted for about 3 weeks. The systemic effect of crude or methanolic neem extract was also observed against the leafminer Liriomyza trifolii (Burgess) by drenching the soil (78,109).

The effects of neem derivatives on stored grain insects have been reviewed indepth (Saxena et al., In The Neem Tree; Jacobson, M., Ed.; State-of-the-Art Series in Phytochemical Pesticides, Vol. 1, CRC, Boca Raton, in press,) and this topic needs no elaboration. Neem derivatives such as dried leaves, seed kernel powder, cake, oil, and others, when mixed with grains in storage, not only protected them from damage, but also caused inhibition of insect growth and development and adult emergence.

Zebitz (44) reported growth inhibitory effects of neem seed extracts and azadirachtin on three species of mosquitoes. Continuous exposure of first instar larvae to neem-treated water delayed the development. Most larvae died as intermediates during imaginal development, similar to the effect of the JH analog Altosid. The growth inhibitory effect was distinct in semi-field trials with Culex pipiens L.. Topical application or injection of azadirachtin into blue blowfly Calliphora vicina (R.-D.) delayed pupation, and decreased pupal weight and adult emergence. Emerging flies were small and malformed.

Azadirachtin possibly inhibits ecdysis by disrupting the ecdysterone titre in insects or by disturbing the neuro-endocrine system, prothoracicotropic hormone, and allatotropic hormone, which control the titres of molting hormone and juvenile hormone, respectively (110). New neem compounds, deacetylazadirachtinol (23,110) and vepaol (28), also inhibit ecdysis.

Raguraman et al. (Int. Rice Res. Newsl., in press) reported that neem derivatives impaired the development of N. lugens nymphs by acting as antagonists of the yeast-like symbionts (YLS) harbored by the insect. The population of YLS was significantly lower in individuals that fed on neem-treated rice plants than on untreated plants. The decrease in number of YLS in nymphs exposed to neem derivatives was either due to the neem's direct antibiotic effect on endosymbionts or to the unfavorable environment rendered in the fat bodies which are YLS depots. It is known that insect brain and fat bodies are sensitive to azadirachtin (45).

Effect on Survival and Reproduction. Neem derivatives do not kill insects directly, but they eventually succumb to behavioral and physiological stresses and starvation on treated plants (89,111-13). Nevertheless, the insects' reproductive physiology is invariably impaired.

Females of E. varivestis (100,114) and L. decemlineata (115,116) were sterilized when exposed to neem components. Production of eggs in treated females and hatchability were greatly reduced. The chorion of eggs was poorly formed leading to fungal infection. The structure of ovaries of E. varivestis females was altered when they fed on bean plants sprayed with 1% methanolic neem kernel extract (117). Cell boundaries of trophocytes disappeared, the cytoplasm of the nurse cells became vacuolized, the nuclei became pycnotic, and the nurse tissue became syncytial. Ovulation was rare, the oocytes and the ovaries shrank, and the vitellarium and the oviducts were resorbed. Egg production also decreased in leafhoppers and planthoppers caged on rice plants sprayed with 12% neem oil (53,89). The egg-laying capacity of the queen of Formica polyctena Foerster ant plunged when fed on a neem oil-treated diet (67). Administration of azadirachtin to Dysdercus koenigii (F.) females led to trophocyte damage (118). Likewise, azadirachtin affected ovaries in last instar nymphs of the milkweed bug Oncopeltus fasciatus (Dallas) (119) and in locusts (120).

Saxena & Barrion (121,122) reported that, compared to the controls, frequencies of meiotic cells were significantly less in male progenies of N. lugens and N. virescens collected from rice plants sprayed with neem seed kernel extract. Few spermatogonia carried on the meiotic divisions that produced primary and secondary spermatocytes, resulting in lower meiotic indices. Azadirachtin suppressed the spermiogenesis of the diapausing pupae of the cabbage armyworm *Mamestra brassicae* (L.) and caused spermatocytes to degenerate in the testicular epithelium (123).

Effect on Mating, Oviposition and Hatchability of Eggs. Little is known about the effect of neem derivatives on insect mating. Fecundity of *N. virescens* was reduced when females were sprayed with neem oil and paired with males on untreated plants (Heyde, J. v. d.; Saxena, R. C.; Schmutterer, H., unpublished data.) Topical application of neem oil on *N. lugens* females at 2.5 or 5 ug/individual or caging them on rice plants sprayed with >3% neem oil disrupted normal courtship signal production and mating behavior (Table II) (Saxena

Table II. Effect of neem oil on courtship signals and mating behavior of *N. lugens* females (Saxena et al. *Int. Rice Res. Newsl.*, in press)[a]

Treatment	PRF of[b] signal (HZ)	Duration[c] of female call (sec)	Premating[d] period (sec)	Duration[e] of mating (sec)
Topical (µg/female)				
1.0	21.5 b	75.9 a	113.5 a	66.5 a
2.5	21.9 b	38.7 b	230.0 ab	72.8 a
5.0	20.3 ab	14.8 c	314.6 b	65.0 a
0 (control)	18.7 a	84.2 a	89.8 a	76.9 a
Foliar (%)				
3	20.8 b	52.1 b	91.2 a	64.2 a
6	20.8 b	27.0 bc	268.0 b	66.7 a
9	21.2 b	20.2 c	219.5 ab	55.5 a
0 (control)	19.1 a	82.8 a	94.8 a	65.5 a

[a]In a column, means followed by a common letter are not significantly different at 5% level by Duncan's multiple range test; [b-e]averages of 8, 36, 12, and 10 replications, respectively.
PRF = pulse repetition frequency of signal.

et al., *Int. Rice Res. Newsl.*, in press.). At higher concentrations of neem oil, most females did not emit signals; consequently, the males could not locate them. Although a few treated females emitted normal signals (at higher pulse repetition frequency), the duration of each call decreased significantly. Unlike the calls of normal males and females, the calls of treated females were not in tune with male calls.

Neem oil deterred egg-laying by homopterans, such as *N. lugens* (89), *Amrasca devastans* (Distant) (55), and *Bemisia tabaci* (Gennadius) (58). Oviposition as well as hatchability were reduced on neem oil-treated substrates in several lepidopterans such as *C. medinalis* (90), *M. separata*, *S. m. acronynctoides* (31,32), *S. litura* (124), *H. armigera* (124), and potato tuber moth *Phthorimaea operculella* (Zell.) (35). Even the vapor of neem oil deterred the production of viable eggs in *Corcyra cephalonica* (Stain.) and *Earias fabia* Stoll (71,73). Oviposition by *N. virescens* and the hatchability of eggs decreased on rice seedlings which had been systemically treated with neem kernel extract for 24 h (126).

Methods of Application

Neem derivatives have been applied against several species of storage pests and crop pests as leaves, oil, cake, extracts, and even as electrodyn formulations of neem oil. They have been used either alone or in selected combination with other edible oils, insecticides, and synergistic compounds. Against crop pests, neem derivatives have been applied using high volume, low volume or ultra-low volume spray applicators, and as presowing treatments, soil applications, or seedling root-dip treatment.

A potentiation of insecticidal activity against *N. virescens* was demonstrated in mixtures of custard-apple oil and neem oil (112), in neem kernel or cake-mixed carbofuran (127), and in combined treatment with neem extract and *Bacillus thuringiensis* (Berliner) (128). The effectiveness of a neem seed extract increased against *P. xylostella* larvae by the addition of a synergist piperonyl butoxide (129). However, the synergist did not stabilize the extract against degradation by ultraviolet rays. Degradation of neem oil by sunlight was reduced by adding 1% carbon or 2% liquid latex (31,32).

Field Trials

Several field trials using neem derivatives for the control of major insect pests of rice and virus diseases transmitted by them have yielded useful data on pest and disease reduction and have increased yields (50,89,130). At the IRRI experimental farm, effective neem treatments such as weekly spraying of 50% neem oil-custard-apple oil mixture in 4:1 proportion (vol/vol) at 8 liters/ha from the seedling to the maximum tillering stage not only decreased tungro disease incidence but also increased the yield (130). The low input cost of the treatment contributed to a high net gain when compared with the use of a recommended insecticide (131).

In field trials in India in 1984-85, neem treatments were found effective against populations of N. virescens, the yellow stem borer Scirpophaga incertulas (Walker), the rice gall midge Orseolia oryzae (Wood-Mason), and grasshoppers (50). Plots sprayed with 2% neem seed extract at 10 kg/ha yielded the highest grain yield.

Neem derivatives also proved effective against H. armigera on Bengal gram, Cicer arietinum (L.) (132, 133), against leafroller Sylepta derogata (F.) and flea beetles Podagrica spp. on okra (83), and against the pod borers H. armigera, Maruca testulalis (Geyer), and Melanagromyza obtusa (Mall.) on pigeon pea (134). In research and commercial greenhouse tests in the USA, a 0.4% crude neem extract applied as a soil drench caused significant mortality of late instars and pupae of the leafminer L. trifolii on chrysanthemums (109). The use of a commercial formulation of soil-applied neem has been suggested for other floricultural crops.

At farm level storage and warehouses, the application of neem derivatives to bags and stored grains has provided protection against insect pests. Powdered neem seed kernel mixed with paddy (1 to 2%) significantly reduced pest infestation in warehouses (17). Neem leaves mixed with paddy (2%), bags treated with 2% neem extract, or a 20-to 30-cm-thick dried neem leaf barrier between the bags and storage floor significantly reduced insect infestation and damage to the grain during a 3-month storage period (135). The effectiveness of neem treatments was comparable to 2% methacrifos dust. Likewise, neem seed extract at 7.2 g/90 kg capacity jute bag (100 x 60 cm) controlled 80% of the population of major insects and checked the damage to stored wheat up to 6 months (136). The treatment was effective up to 13 months and provided more than 70% protection as compared with untreated control. The neem seed extract treatment was as effective as that of 0.0005% primiphos methyl mixed with the grain. Using this technology in Sind, Pakistan, high benefit-cost ratios were attained by small- (4.6), medium- (5.6), and large-scale (7.4) farmers (137).

Reduction of Virus Transmission

Simons (138) emphasized the use of oil formulations with antifeedants from plants for controlling insect-transmitted viral diseases. Antifeedants, though not toxic, prevent or reduce insect feeding. Saxena et al. (89) observed that the incidence of *N. lugens*-transmitted ragged stunt virus disease was significantly less in ricefields periodically sprayed with 12% neem oil than in unsprayed field. Neem oil reduced the *N. lugens* survival and suppressed the transmission of grassy- and ragged stunt virus diseases (113). Neem oil alone or mixed with custard-apple oil also reduced the survival of *N. virescens* and its transmission of rice tungro viruses on treated seedlings (111,112). A 5% neem cake extract or 2% kernel extract significantly reduced the incidence of yellow mosaic virus transmitted by the whitefly *B. tabaci* in black gram (139).

The incidence of virus diseases was generally lower in ricefields applied with a neem cake-urea mixture than in fields applied with urea only (91). Enzyme-linked immunosorbent assays showed that rice seedlings grown in soil treated with neem cake at 150 kg/ha had a significantly lower incidence of infection of rice tungro bacilliform and spherical viruses transmitted by *N. virescens* than in control seedlings (96). Protection with neem cake at 250 kg/ha was comparable with that provided by application of carbofuran 3G at 0.75 kg (AI)/ha.

Safety of Neem to Beneficial Organisms and Man

Saxena et al. (90) recorded that parasitization of *C. medinalis* larvae was higher in ricefields sprayed with 50% neem oil than in unsprayed fields. Normal male and female *Macrocentrus philippinensis* Ashmead emerged from parasitized *C. medinalis* larvae which had been confined to rice leafcuts treated with partially-purified neem seed fractions (107). Likewise, neem seed kernel suspension in water sprayed on *Spodoptera litura* Fab. eggs before or after parasitization did not affect the emergence of the egg parasite *Telenomus remus* Nixon (140).

The mirid *Cyrtorhinus lividipennis* (Reuter) and the wolf spider *Lycosa pseudoannulata* (Boesberger & Strand), predators of rice insects, were not affected by topically applied neem oil at doses of up to 20 µg/adult and 50 µg/adult, respectively (92). *N. lugens* population buildup was less on plants sprayed with a solution of neem seed bitters than on control plants with or without the predatory mirid; pest population was least when the predator was also introduced on neem-treated plants (36). In field trials, neem treatments did not affect mirids and spiders preying on rice insects (141). Neem seed extracts were considerably toxic to the phytophagous mite *Tetranychus cinnabarinus* (Boisd.) but not to its predaceous mite *Phytoseiulus parsimilis* Athias-Henriot and the spider *Chiracanthium mildei* L. Koch (142).

Ground neem leaves or seed kernel incorporated in potted soil containing the earthworm *Eisenia foetida* Savigny increased the number of young worms by 25% more (143). A purified neem extract up to 100 ppm did not harm young guppies *Poecilia reticulata* (= *Lebistes reticulatus*) Peters (44). Incorporation of 20% neem cake in sheep diet increased the growth rate (144). Rats fed with a neem extract at a dose up to 600 mg/kg body weight increased body weight without overt toxicity (145). Neem kernel extracts did not inhibit spermatogenesis in the rat (146,147). In the standard Ames tests, azadirachtin showed no mutagenic activity on strains of *Salmonella typhimurium* (Loeffler) Castellani-Chalmers (22). Normal human cells in culture were not affected by neem extracts at 5 mg/ml, while tumor-originated cells were degenerated (148). Recently, a neem oil-based, precoital, intravaginal, spermicidal contraceptive has been made in India.

Resistance of Insects to Neem Derivatives?

A predictable response to insecticide use is the evolution of resistant populations of insect pests. Taylor (149) indicated that insects may possibly adapt to limonoids rather quickly. However, Völlinger (150) demonstrated that two genetically different strains of *P. xylostella* treated with a neem seed extract showed no sign of resistance in feeding and fecundity tests up to 35 generations. In contrast, deltamethrin-treated lines developed resistance factors of 20 in one line and 35 in the other. There was no cross resistance between deltamethrin and neem seed kernel extract in the deltamethrin-resistant lines. The activity of esterase and multi-function oxidase enzymes did not change during the 35 generations.

Scope and Prospects of Neem-based Insecticides

Neem has had a long history of use primarily against household and storage pests in the Indian subcontinent where it originated. The present review reveals that neem derivatives affect nearly 200 species of insects. Attention was earlier paid only to the linkage between the neem's bitterness and insect repellence and phagodeterrent properties. Recently, other behavioral and physiological effects of neem derivatives have been unravelled, providing new opportunities for insect pest control. However, sustained and concerted efforts are needed to avail of the full potential of neem as a source for "insecticides."

Although 280 limonoids, including azadirachtin,

from Meliaceous plants have been identified (149), more may be discovered. Yet their complexity will preclude their chemical synthesis in the near future. The alternative approach of obtaining bioactive fractions from neem seed is therefore feasible and economically viable. In this context, advances made in extracting the bioactive components from neem using inexpensive solvents are promising and may facilitate large-scale processing of neem seeds during the harvest season. The use of neem bitters and other bioactive fractions, rather than neem oil as such will ease the competition with the soap industry, which at present is a major buyer of neem oil. The semipurified neem seed bitters and other neem-rich fractions can easily be standardized for biological properties and can satisfy even stringent quality requirements. Being water-soluble, they can also be applied as systemic compounds, which renders them more photostable and nonphytotoxic, unlike the neem oil. Their efficacy can further be enhanced with additives, synergists, and antioxidants, and improved methods of application.

More neem trees will have to be grown to meet the increasing demand for industrial and insecticidal uses. Selection of superior ecotypes will be desirable to increase the productivity and quality of neem produce. Employing tissue culture and biotechnological methods, it may even be possible to develop cold-tolerant ecotypes of neem, which is otherwise a tropical tree. The underutilized neem lots in some African countries such as Sudan, Nigeria, Togo, Ghana, and others can be tapped and will open new opportunities for employment and income generation. The hardy, evergreen tree can also be propagated in degraded soils and denuded environments. Unlike Pyrethrum, once established, neem trees will be good sources of raw materials for industrial and insecticidal use even for as long as 300 years.

The apprehension that large scale use of neem-based inseticides may lead to resistance among insect pests has not been substantiated. Unlike conventional insecticides based on a single active ingredient, the bioactive components in neem comprise a complex array of novel chemicals that affect not only one physiological function but rather act in concert on a number of behavioral and physiological processes. Consequently, chances of insect pests developing resistance to neem materials are remote. Neem-based insecticides can further be fortified against dynamic insect pests by optimizing their use with microbials, such as *B. thuringiensis*, nuclear polyhedroses viruses, entomophagous pathogens, synthetic insecticides, or plant derivatives.

The current void caused by nonavailability of

selective chemicals that affect the pests but spare the natural enemies in insect pest management programs can be filled by neem-based insecticides or even by the use of crude extracts or derivatives of neem. The safety of neem derivatives to predators and parasites and other nontarget organisms, including man, is now well established. Still it would require a revision of the current policies of qualifying insecticides based on chemical structures rather than on biological activity.

An enlightened attitude toward pest control philosophy would be helpful in promoting neem-based insecticides. Killing and destruction of pests is not necessary if they can be incapacitated otherwise. In this context, the use of neem derivatives offers a harmonious and nonviolent approach to pest management. Also, the neem compounds' ability to inhibit nitrification, should make their use in crop husbandry doubly attractive.

To benefit fully, the cost of neem-based insecticides must remain within the user's reach. To achieve this, farmers should be encouraged to grow neem on their homesteads and be educated about its values in insect control.

Acknowledgment

Most of the neem research carried out at IRRI was made possible through financial grants from The Asian Development Bank, Manila, Philippines, and The Directorate for Technical Cooperation and Humanitarian Aid, Switzerland. I thank A. Abdul Kareem, visiting scientist at IRRI, for useful comments and help in the preparation of this review.

Literature Cited

1. Mandava, N. B. Handbook of Natural Pesticides: Methods, Theory, Practice, and Detection, CRC, Boca Raton, Florida, 1985, p 534.
2. Pimentel, D.; Krummel, J.; Gallahan, D.; Hough, J.; Merrill, A.; Schreinder, I.; Vittum, P.; Koziol, F.; Back, E.; Yen, D.; Fiance, S. BioScience 1978, 28, 772, 778-784.
3. Anonymous. In Chemistry and World Food Supplies: The New Frontiers, CHEMRAWN II, Perspectives and Recommendations; Bixler, G.; Shemilt, L. W., Eds.; IRRI, Manila, Phil., 1987, pp 5-23.
4. Williams, C. M. Sci. Am. 1967, 217, 13-17.
5. Bowers, W. S. In Pesticide Chemistry in the 20th Century; Plimmer, J. R., Ed.; ACS Symposium Series No. 37, Washington, DC, 1977; pp 271-275.
6. Casida, J. E. In Pesticide Science and Biotech-

nology; Greenhalgh, R.; Roberts, T. R., Eds.; Blackwell Scientific, London, 1987; pp 75-80.

7. Schooley, D.; Quistad, G. B.; Skinner, W. S.; Adams, M. E. In Pesticide Science and Biotechnology; Greenhalgh, R.; Roberts, T. R. Eds.; Blackwell Scientific, London, 1987; pp 97-100.
8. Fredenhagen, A.; Kenny, P.; Kita, H.; Komura, H.; Naya, Y.; Nakanishi, K.; Nishiyama, K.; Sigiura, M.; Tamura, S. In Pesticide Science and Biotechnology; Greenhalgh, R.; Roberts, T. R. Eds.; Blackwell Scientific, London, 1987; pp 101-108.
9. Swain, T. Ann. Rev. Plant Physiol. 1977, 28, 479-501.
10. Roxburgh, W. Description of Indian Plants, Today & Tomorrow, New Delhi, 1874; p 763 (reprinted from Carey's edition 1832.)
11. Watt, G. A Dictionary of the Economic Products of India, Periodical Experts, New Delhi 1972; Vol. 15, p 676 (reprinted from 1891 edition.)
12. Pruthi, H. S.; Singh, M. Stored Grain Pests and Their Control, Imperial Council of Agric. Res., Misc. Bull. No. 57, 1944.
13. Department of Agriculture, Bengal. Department of Agriculture, Bengal (India), 1932, Leaflet No. 2.
14. Pradahan, S.; Jotwani, M. G.; Rai, B. K. Indian Farming 1962, 12, 7-11.
15. Lavie, D.; Jain, M. K.; Shpan-Gabrielith, S. R. J. Chem. Soc. 1967, Chem. Commun. 910-911.
16. Butterworth, J. H.; Morgan, E. D. J. Chem. Soc. 1968, Chem. Commun. 23-24.
17. Ketkar, C. M. Utilization of Neem (Azadirachta indica Juss) and its Bye-products; Directorate of Non-edible Oils & Soap Industry, Khadi & Village Industries Commission, Bombay, 1976; p 234.
18. Schmutterer, H.; Rembold, H.; Ascher, K. R. S. Proc. 1st Int. Neem Conf., Rottach-Egern, 1980, p 297.
19. Schmutterer, H.; Ascher, K. R. S. Proc. 2nd Int. Neem Conf., Rauischholzhausen, 1983, p 587.
20. Schmutterer, H.; Ascher, K. R. S. Proc. 3rd Int. Neem Conf., Nairobi, 1986, p 703.
21. Jacobson, M. Proc. 1st Int. Neem Conf., Rottach-Egern 1980, pp 33-42.
22. Jacobson, M. In Natural Resistance of Plants to Pests: Roles of Allelochemicals; Green, M. B.; Hedin, P. A., Eds.; ACS Symposium Series No. 296; American Chemical Society: Washinton, DC, 1986; pp 220-232.
23. Jacobson, M. Proc. 3rd Int. Neem Conf., Nairobi, 1986, pp 33-44.
24. Kraus, W.; Cramer, R.; Bokel, M.; Sawitzki, G.

Proc. 1st Int. Neem Conf., Rottach-Egern, 1980, pp 53-62.
25. Kraus, W., Baumann, S. ; Bokel, M.; Keller, U.; Klenk, A.; Klingele, M.; Pohnl, H.; Schwinger, M. Proc. 3rd Int. Neem Conf., Nairobi, 1986, pp 111-125.
26. Morgan, E. D. Proc. 2nd Int. Neem Conf., Rottach-Egern, 1980, pp 43-52.
27. Henderson, R.; McCrindle, R.; Overton, K. H. Tetrahedron Lett. 1964, 24, 1517-1523.
28. Sankaram, A. V. B.; Murthy, M. M.; Bhaskaraiah, K.; Subramanyam, M.; Sultana, N.; Sharma, H. C.; Leuschner, K.; Ramaprasad, G.; Sitaramaiah, S.; Rukmini, C.; Rao, P. U. Proc. 3rd Int. Neem Conf., Nairobi, 1986, pp 127-148.
29. Lee, S. M.; Olsen, J. I.; Schweizer, M. P.; Klocke, J. A. Phytochemistry 1988, 27, 2773-2776.
30. Broughton, H. B.; Ley, S. V.; Slawin, A. M. Z.; Williams, D. J.; Morgan, E. D. J. Chem. Soc. 1986, Chem. Commun. 1365, 46.
31. Saxena, R. C. In Pesticide Science and Biotechnology; Greenhalgh, R.; Roberts, T. R., Eds.; Blackwell Scientific, London, 1987; pp 139-144.
32. Saxena, R. C. Proc. 3rd Int. Neem Conf., Nairobi, 1986, pp 81-93.
33. IARI (Indian Agricultural Research Institute). Neem in Agriculture, Res. Bull. No. 40, New Delhi, 1983, p 63.
34. Larson, R. O. Proc. 3rd Int. Neem Conf., Nairobi, 1986, pp 243-250.
35. Sharma, R. N.; Nagasamapagi, B. A.; Bhosale, S. A.; Kulkarni, M. M.; Tungikar, V. B. Proc. 2nd Int. Neem Conf., Rauischholzhausen, 1983, pp 115-128.
36. Saxena, R. C.; Rueda, B. P.; Justo, Jr., H. D.; Boncodin, M. E. M.; Barrion, A. A. Proc. 18th Ann. Conf., Pest Control Council of the Philippines, 1987, pp 1-43.
37. Naqvi, S. N. H. Proc. 3rd Int. Neem Conf., Nairobi, 1986, pp 315-330.
38. Islam, B. N. Proc. 3rd Int. Neem Conf., Nairobi, 1986, pp 217-242.
39. Reed, D. K.; Reed, G. L. Indiana Acad. Sci. 1985, 94, 335-339.
40. Gunathilagaraj, K.; Sundara Babu, P.C. Proc. Workshop on Botanical Pest Control in Rice-based Cropping Systems, Coimbatore, 1987, pp 9.
41. Borah, D.; Saharia, D. J. Res. Assam Agric. Univ., 1985, 3, 224-226.
42. Karel, A. K. Proc. 3rd Int. Neem Conf.,Nairobi, 1986, pp 393-403.

43. Chin, S. L.; Suderuddin, K. I.; Khoo, S. G. Malayasian Applied Biol. 1980, 9, 59-66.
44. Zebitz, C. P. W. Proc. 3rd Int. Neem Conf., Nairobi, 1986, pp 555-573.
45. Bidmon, H.J.; Kauser, G.; Mobus, P; Koolman, J. Proc. 3rd Int. Neem Conf., Nairobi, 1986, pp 253-271.
46. Singh, R. P.; Srivastava, D. G. Indian J. Ent. 1983, 45, 497-498.
47. Srivastava, K.P.; Agnihotri, N. P.; Gajbhiye, V. T.; Jain, H. K. J. Entomol. Res. 1984, 8, 1-4.
48. Warthen Jr., J. D.; Uebel, E. C.; Dutky, S. R.; Lusby, W. R.; Finegold, H. Adult Housefly Feeding Deterrent from Neem Seeds, U.S. Department of Science and Administration, Agricultural Research Results, ARR, NE2, 1978.
49. Rajasekaran, B.; Jayaraj, S.; Ravindran, R. Proc. Workshop on Botanical Pest Control in Rice-based Cropping Systems, Coimbatore, 1987, pp. 9.
50. Saroja, R. Int. Rice Res. Newsl. 1986, 11(4), 33-34.
51. Wilps, H. Proc. 3rd Int. Neem Conf., Nairobi, 1986, pp 299-314.
52. Leuschner, K. Naturwissenscaften 1972, 5, 217-218.
53. Heyde, J. v. d.; Saxena, R. C.; Schmutterer, H. Proc. 2nd Int. Neem Conf., Rauischholzhausen 1983, pp 377-390.
54. Saxena, R. C.; Justo,Jr., H. D.; Domingo, I.; Shepard, B. M.; Aguda, R. M.; Rombach, M. C. Proc. Workshop on Botanical Pest Control in Rice-based Cropping Systems, Coimbatore, 1987, pp 2.
55. Saxena, K. N.; Basit, A. J. Chem. Ecol. 1982, 8, 329-338.
56. Chiu, Shin-Foon; Zeng, Xin-Nian. Neem Newsl. 1986, 3(4), 48.
57. Siddig, S. A. Proc. 3rd Int. Neem Conf., Nairobi, 1986, pp 449-459.
58. Coudriet, D. L.; Prabhakar, N.; Meyerdirk, D. E. Environ. Entomol. 1985, 14, 776-779.
59. Karganilla, G. S. Proc. 3rd Int. Neem Conf., Nairobi, 1986, (abstract.)
60. Dreyer, M. Proc. 3rd Int. Neem Conf. Nairobi, 1986, pp 431-447.
61. Sharma, R. K.; Saxena, R.; Naithani, S. Neem Newsl. 1986, 3(1), 1-2.
62. Srivastava, K. P.; Parmar, B. S. Neem Newsl. 1985, 2(1), 7.
63. Griffiths, D. C.; Greenway, A. R.; Lloyd, S. L. Bull. Ent. Res. 1978, 68, 613-619.
64. Goyal, R. S.; Gulati, K. C.; Sarup, P.; Kidwai, M. A.; Singh, D. S. Indian J. Entomol. 1972, 33, 67-71.

65. Panda, N. Proc. Workshop on Botanical Pest Control in Rice-based Cropping Systems, Coimbatore, 1987, pp 12-13.
66. Larew, H. G.; Knodel-Montz, J. J.; Marion, D. F. J. Environ. Hort. 1987, 5, 17-19.
67. Schmidt, G. H.; Pesel, E. Proc. 3rd Int. Neem Conf., Nairobi, 1986, pp 361-373.
68. Chari, M. S.; Muralidharan, C. M. J. Entomol. Res. 1985, 9, 243-245.
69. Saxena, R. C. Indian J. Agric. Sci. 1982, 52, 51-52.
70. Meisner, J.; Wysoki, M.; Ascher, K. R. S. Phytoparasitica 1976, 4, 185-192.
71. Pathak, P. H.; Krishna, S. S. Z. Ang. Ent. 1985, 100, 33-35.
72. Fagoonee, I. Reveree Agricale et Sucriere de I'lle Maurice 1980, 59, 57-62.
73. Pathak, P. H.; Krishna, S. S. Appl. Ent. Zool. 1986, 21, 347-348.
74. Parmar, B. S.; Srivastava, K. P. Proc. 3rd Int. Neem Conf., Nairobi, 1986, 205-215.
75. Singh, R. P. Proc. Workshop on Botanical Pest Control in Rice-based Cropping Systems, Coimbatore, 1987, pp 14.
76. Ho, D. T.; Kibuka, J. G. Int. Rice Res. Newsl. 1983, 8(5), 15-16.
77. Chiu, Shin-Foon; Huang, Zhang-Xin; Liu, S. K.; Huang, D. P. Acta Ent. Sinica 1984, 27, 241-247.
78. Meisner, J.; Melamed-Madjar, V.; Yathom, S.; Ascher, K. R. S. Proc. 3rd Int. Neem Conf., Nairobi, 1986, pp 461-477.
79. Batra, R.C.; Sandhu, G. S. Pesticides 1981, 15(2), 5-6.
80. Schlütter, U. Proc. 3rd Int. Neem Conf., Nairobi, 1986, pp 331-348.
81. Zhang, Ye-Guang; Chiu, Shin-Foon. Neem Newsl. 1985, 2(3), 30-32.
82. Lidert, Z.; Taylor, D. A. H.; Thirugnaman, M. J. Natural Prod. 1985, 48, 843-845.
83. Adhikary, S. Z. Ang. Ent. 1984, 98, 327-331.
84. Adler, V. E.; Uebel, E. C. Phytoparasitica 1985, 13, 3-8.
85. Olaifa, J. I.; Akinbohungbe, A. E. Proc. 3rd Int. Neem Conf., Nairobi, 1986, pp 405-418.
86. Chiu, Shin-Foon; Huang, Zhang-Xin; Huang, Bin-Qiu; Hu, Mei-Yin. Proc. Workshop on Botanical Pest Control in Rice-based Cropping Systems, Coimbatore, 1987, pp 6.
87. Grant, I. F.; Schmutterer, H. Proc. 3rd Int. Neem Conf., Nairobi, 1986, pp 591-607.
88. Dethier, V. G.; Browne Bartoon, L.; Smith, C. N. J. Econ. Entomol. 1960, 53, 134-136.

89. Saxena, R. C.; Liquido, N. J.; Justo, Jr., H. D. Proc. 1st Int. Neem Conf., Rottach-Egern, 1980, pp 171-188.
90. Saxena, R. C.; Waldbauer, G. P.; Liquido, N. J.; Puma, B. C. Proc. 1st Int. Neem Conf., Rottach-Egern, 1980, pp 198-204.
91. Saxena, R. C.; Justo, Jr., H. D.; Epino, P. B. J. Econ. Entomol. 1984, 77, 502-507.
92. Saxena, R. C.; Epino, P. B.; Tu, Chen-Wen; Puma, B. C. Proc. 2nd Int. Neem Conf., Rauischholzhausen, 1983, pp 403-412.
93. Saxena, R. C.; Khan, Z. R. J. Econ. Entomol. 1985, 78, 222-226.
94. Saxena, R. C.; Khan, Z. R. Entomol. Exp. Appl. 1986, 42, 279-284.
95. Gill, J. S.; Lewis, C. T. Nature 1971, 232, 402-403.
96. Saxena, R. C.; Khan, Z. R.; Bajet, N. B. J. Econ. Entomol. 1987, 80, 1079-1082.
97. Schoonhoven, L. M. Ent. Exp. Appl. 1982, 31, 57-69.
98. Schoonhoven, L. M. In Pesticide Science and Biotechnology, Greenhalgh, R.; Roberts, T. R., Eds.; Blackwell Scientific, London, 1987, pp 129-132.
99. Steets, R. Z. Ang. Ent. 1976, 77, 306-312.
100. Steets, R. Ph. D. Thesis, Giessen University, FRG, 1976.
101. Ruscoe, C. N. E. Nature 1972, 236, 159-160.
102. Rembold, H.; Sharma, G. K.; Czoppelt, Ch.; Schmutterer, H. Z. PflKrankh. PflSchutz 1980, 87, 290-297.
103. Rembold, H.; Sharma, G. K.; Czoppelt, Ch. Proc. 1st Int. Neem Conf., Rottach-Egern, 1980, pp 121-128.
104. Schmutterer, H.; Rembold, H. Z. Ang. Ent. 1980, 89, 179-188.
105. Schlüter, U. Proc. 1st Int. Neem Conf., Rottach-Egern, 1980, pp 97-104.
106. Kubo, K.; Klocke, J. A. Agric. Biol. Chem. 1982, 46, 1951-1953.
107. Schmutterer, H.; Saxena, R. C.; Heyde, J. v. d. Z. Ang. Ent. 1983, 95, 230-237.
108. Heyde, J. v. d.; Saxena, R. C.; Schmutterer, H. Z. PflKrankh. PflSchutz 1985, 92, 346-354.
109. Larew, H. G.; Knodel-Montz, J. J.; Webb, R. E.; Warthen, J. D. J. Econ. Entomol. 1985, 78, 80-84.
110. Kubo, K.; Klocke, J. A. In Natural Resistance of Plants to Pests: Roles of Allelochemicals; Green, M. B.; Hedin, P. A., Eds.; ACS Symposium Series No. 296, American Chemical Society, Washinton, DC, 1986; pp 206-219.

111. Mariappan, V.; Saxena, R. C. J. Econ. Entomol. 1983, 76, 573-576.
112. Mariappan, V.; Saxena, R. C. J. Econ. Entomol. 1984, 77, 519-521.
113. Saxena, R. C.; Khan, Z. R. J. Econ. Entomol. 1985, 78, 647-651.
114. Steets, R.; Schmutterer, H. Z. PflKrankh. PflSchutz 1975, 82, 176-179.
115. Steets, R. Z. Ang. Ent. 1976, 82, 169-176.
116. Schmutterer, H. Proc. 3rd Int. Neem Conf., Nairobi, 1986, pp 351-360.
117. Schulz, W. D. Proc. 1st Int. Neem Conf., Rottach-Egern, 1980, pp 81-96.
118. Koul, O. Z. Ang. Ent. 1984, 98, 221-223.
119. Dorn, A.; Rademacher, J. M.; Sehn, E. Proc. 3rd Int. Neem Conf., Nairobi, 1986, pp 273-288.
120. Rembold, H.; Uhl, M.; Muller, Th. Proc. 3rd Int. Neem Conf., Nairobi, 1986, pp 289-298.
121. Saxena, R. C.; Barrion, A. A. Int. Rice Res. Newsl. 1987, 12(5), 24-25.
122. Saxena, R. C.; Barrion, A. A. Int. Rice Res. Newsl. 1987, 12(5), 25-26.
123. Shimizu, T. Entomol. Exp. Appl. 1988, 46, 197-199.
124. Joshi, B. G.; Sitaramaiah, S. Phytoparasitica 1979, 7, 199-202.
125. Saxena, K. N.; Rembold, H. Proc. 2nd Int. Neem Conf., Rauischholzhausen, 1983, pp 199-210.
126. Abdul Kareem, A.; Saxena, R. C.; Boncodin, M. E. M. Neem Newsl. 5(1), 9-10.
127. Abdul Kareem, A.; Boncodin, M. E. M.; Saxena, R. C. Int. Rice Res. Newsl. 1988, 13(3),35.
128. Hellpap, C. Proc. 2nd Int. Neem Conf., Rauischholzhausen, 1983, pp 353-364.
129. Lange, W.; Feuerhake, K. Z. Ang. Ent. 1984, 98, 368-375.
130. Saxena, R. C.; Justo, Jr., H. D. Int. Rice Res. Newsl. 1986 11(2), 25.
131. Abdul Kareem, A.; Saxena, R. C.; Justo, Jr., H. D. Int. Rice Res. Newsl. 1987, 12(4), 28-29.
132. Kumar, A. R. V.; Sengapa, H. K. Current Research 1984, 13(4/6), 38-40.
133. Siddappaji, A. R.; Kumar, V.; Gangadharaiah. Pesticides 1966, 20(1), 13-16.
134. Singh, R. P.; Singh, J.; Singh, S. P. Indian J. Entomol. 1985, 47, 111-112.
135. Muda, A. R. In ASEAN Crops Postharvest Programme; Sample, R. L.; Frio, A. S., Eds; ASEAN Food Handling Bureau, Manila, Phil., 1984.
136. Jilani, G. Proc. 17th Ann. Conf. Pest Control Council, Philippines, 1986.
137. Jilani, G.; Amir, P. Economics of Neem in Reducing Wheat Storage Losses: Policy

Implications, Tech. Bull. No. 2, SEARCA, Phil., 1987, p 15.

138. Simons, J. N. In Vectors of Disease Agents, Interaction with Plants, Animal and Man, McKelvy, Jr., J. K.; Elridge, B. F.; Maramorosch, K., Eds.; Praeger, N.Y., 1981, pp 169-178.

139. Mariappan, V.; Gopalan, M.; Narasimhan, V.; Suresh, S. Neem Newsl. 1987, 4(1), 9-10.

140. Joshi, B. G.; Ramaprasad, G; Sitaramaiah, S. Phytoparasitica 1982, 10, 61-63.

141. Abdul Kareem, A.; Saxena, R. C.; Boncodin, M. E. M.; Palanginan, E. L.; Malayba, M. T.; Barrion, A. A. Proc. 19th Ann. Conf. Pest Control Council, Philippines, 1988; p 35.

142. Mansour, F.; Ascher, K. R. S.; Omari, N. Proc. 3rd Int. Neem Conf., Nairobi, 1986, pp 577-587.

143. Rossner, J.; Zebitz, C. P. W. Proc. 3rd Int. Neem Conf., Nairobi, 1986, pp 611-621.

144. Vijjan, V. K.; Tripathi, H. C.; Parihar, N. S. J. Environ. Biol. 1982, 3, 47-52.

145. Qadri, S. S. H.; Usha, G.; Jabeen, K. Int. Pest Control 1984, 26, 18-20.

146. Krause, W.; Adami, M. Proc. 2nd Int. Neem Conf., Rauischholhausen, 1983, 483-490.

147. Sadre, N. L.; Deshpande, V. Y.; Mendulkar, K. N.; Nandal, D. H. Proc. 2nd Int. Neem Conf., Rauischholzhausen, 1983, pp 473-482.

148. Gogate, S. S.; Marathe, A. D. Indian J. Microbiol. 1986, 26, 185-186.

149. Taylor, D. A. H. In Progress in the Chemistry of Organic Natural Products, Herz, W.; Grisebach, H; Kirby, G. W., Eds; Springer-Verlag, Vienna, 1984; Vol. 45, pp 1-102.

150. Völlinger, M. Proc. 3rd Int. Neem Conf., Nairobi, 1986, pp 543-554.

RECEIVED November 18, 1988

Chapter 10

Limonoids, Phenolics, and Furanocoumarins as Insect Antifeedants, Repellents, and Growth Inhibitory Compounds

James A. Klocke, Manuel F. Balandrin, Mark A. Barnby, and R. Bryan Yamasaki

Native Plants, Inc. (NPI), University of Utah Research Park, Salt Lake City, UT 84108

Plants biosynthesize a dazzling array of structural types which exhibit an almost equally dazzling array of biological activities. In insects, various plant compounds affect nerve axons and synapses (e.g., pyrethrins, nicotine, picrotoxinin), muscles (e.g., ryanodine), respiration (e.g., rotenone, mammein), hormonal balance (e.g., juvenile and molting hormone analogs and antagonists), reproduction (e.g., β-asarone), and behavior (e.g., attractants, repellents, antifeedants). Some of these compounds have already been exploited in commercial insect control (e.g., pyrethrins, juvenile hormone analogs, attractants), while others offer a unique opportunity as sources and models of new insect control agents (1,2). Still others may be important components of host plant resistance mechanisms. This chapter will focus on the activity of certain plant compounds as insect antifeedant, repellent, and growth inhibitory compounds.

Insect Antifeedants

Antifeedants are substances which, when tasted by insects, result either temporarily or permanently, depending on potency, in the cessation of feeding (3). The existence of and potential for antifeedant compounds, both natural and synthetic, in practical insect control have been known for some time. For example, Bordeaux mixture (copper sulfate, hydrated lime, and water), known for over 100 years, acts as a feeding deterrent to flea beetles, leaf hoppers, and the potato psyllid (*Paratrioza cockerelli* (Psyllidae)) (4). In the 1930's, a number of field trials were carried out in the U.S. with insect antifeedant compounds, including some derived from plants (5). Success was limited and was not sufficient to compete with synthetic insecticides. Only more recently have the problems associated with total reliance on synthetic pesticides necessitated the reevaluation of antifeedant compounds.

0097–6156/89/0387–0136$06.00/0

Several synthetic compounds, including 4'-(dimethyltriazeno)acetanilide (a triazene) and triphenyltin acetate (an organotin), have been tested in field trials for insect antifeedant activity (6, 7). Organotin compounds are presently used as antifeedants on several crops in Africa (8). Some commercial insecticides also exhibit antifeedant activity (e.g., agricultural pyrethroids) (9).

Several groups in the last decade have been examining plant extracts and pure plant compounds for insect antifeedant activity (3, 10-12). Although none of the plant compounds have thus far been developed as commercial products, several of the more active ones have been synthesized in the hope of making these compounds, and simpler structural analogs, more widely available for testing (13). For example, several drimane sesquiterpenoids (including warburganal and polygodial) which were shown to have potent activity against the African armyworm, *Spodoptera exempta* (Noctuidae), have been synthesized (13, 14). Other insect antifeedant plant compounds have also been synthesized (13), including some clerodane diterpenoids (e.g., clerodin and ajugarin I) found to be antifeedants for a number of species of insects (10, 15). A precise arrangement of the functional groups of the clerodane diterpenoids, including the *trans*-epoxydiacetate and the furofuran or butenolide-containing side chain, was found necessary for antifeedant activity (13, 16).

For several reasons these compounds have not yet been commercialized. For example, warburganal is a potent antifeedant against *S. exempta*, but has little activity against most other insect species tested (17). In addition, the strong cytotoxicity and hemolytic properties of warburganal may preclude its practical use (17). In the case of the ajugarins, they may not be potent enough to warrant commercialization as insect antifeedants (I. Kubo, Univ. of California, Berkeley, personal communication). As with any insect control agents, cost is a major consideration. Antifeedants are no less costly to use than are conventional insecticides (5) and they require the same exhaustive study to prove their toxicological safety (8).

Several requirements should be met before a compound can be effectively used as an antifeedant in commercial insect control. For example, in the field, insects will choose untreated plant parts over those treated with antifeedant compounds. It is therefore desirable that a candidate antifeedant be taken up systemically by the plant to assure complete protection.

On the other hand, starving insects will often attempt feeding on any potential food available to them, including that treated with antifeedant compounds. In this case, the compounds would be acting as "relative" antifeedants, that is, they inhibit feeding only for a defined time (10). Although a number of plant compounds exhibit relative antifeedant activity, only a few exhibit "absolute" antifeedant activity such that the insects die from starvation rather than start eating treated foodstuffs. It is therefore desirable that a candidate antifeedant also have some toxic action if ingested.

One natural plant product that may fulfill these criteria is the limonoid azadirachtin (Figure 1) (see other chapters in this

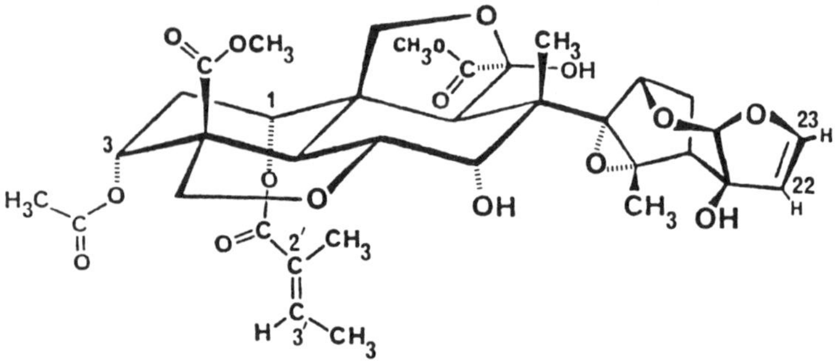

AZADIRACHTIN

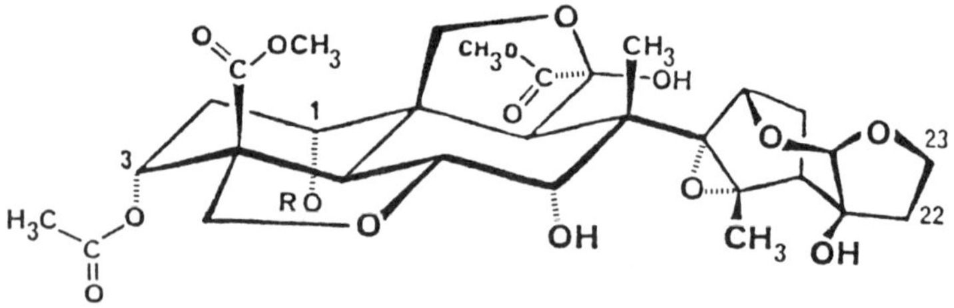

22,23-DIHYDROAZADIRACHTIN; R = $\overset{H_3C}{\underset{3'}{HC}}=\overset{CH_3}{\underset{2'}{C}}CO$

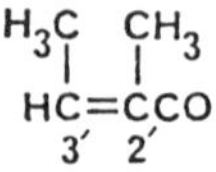

2′,3′,22,23-TETRAHYDROAZADIRACHTIN; R = $CH_3\overset{3'}{CH_2}\underset{CH_3}{\overset{2'}{CH}}CO$

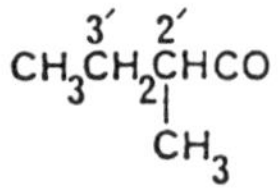

Figure 1. Stereostructures of three insect antifeedant limonoids.

volume). The antifeedant activity of azadirachtin, and the extracts containing it, is well documented (18, 19). For example, against the fall armyworm, Spodoptera frugiperda (Noctuidae), 0.3 $\mu g/cm^2$ of azadirachtin resulted in <5% of treated cotton leaf disks being eaten, while >95% of untreated disks were eaten in "choice" assays (Table I). In field studies conducted in our collaboration with a major U.S. chemical company, azadirachtin was found to be systemic in corn and to be effective as a prophylactic against S. frugiperda at 0.5 lb/ha.

Azadirachtin is less potent against some other species of insects. For example, against the Colorado potato beetle, Leptinotarsa decemlineata (Chrysomelidae), 60 $\mu g/cm^2$ of azadirachtin resulted in <5% of treated potato leaf disks being eaten, while >50% of untreated disks were eaten in "choice" assays (Table I).

Two derivatives of azadirachtin, 22,23-dihydroazadirachtin and 2',3',22,23-tetrahydroazadirachtin (Figure 1), were prepared according to Yamasaki and Klocke (20) and tested as antifeedants against S. frugiperda and L. decemlineata. Similar to our results with these compounds as growth inhibitors and toxicants to the tobacco budworm, Heliothis virescens (Noctuidae) (20), these two derivatives were as active as azadirachtin itself as antifeedants against S. frugiperda (Table I). However, as antifeedants against L. decemlineata, the two derivatives were less active than azadirachtin (Table I). Therefore, structure-bioactivity relationship (SAR) studies involving azadirachtin as an insect antifeedant should be conducted with several species of insects since the response of each species to individual derivatives may differ.

Some attempts have been made to define the SAR of azadirachtin as an antifeedant. For example, the furopyran moiety of azadirachtin has been suspected by some to be the active center responsible for the antifeedant activity (21). However, Pflieger and Muckensturm (21) found no antifeedant activity with 9-hydroxydihydrofuro-2,3-pyran against the Egyptian cotton leafworm, Spodoptera littoralis (Noctuidae). On the other hand, Ley et al. (22) reported that a novel hydroxydihydrofuran acetal, which represents a fragment of azadirachtin, is nearly as potent as azadirachtin as an antifeedant against S. littoralis.

In addition to its antifeedant effects, azadirachtin is a slow acting insecticide because it disrupts the hormonal balance in certain species of insects when ingested (1,23). Preliminary data in our laboratory indicate that azadirachtin inhibits the release of insect brain hormone (prothoracicotropic hormone).

Insect Ovipositional Repellents

Our interest in insect ovipositional repellents stems from recent investigations in our laboratory on the biologically active volatile oil of the tarweed, Hemizonia fitchii (Asteraceae) (24). From H. fitchii volatile oil, we isolated the monoterpenoid, 1,8-cineole (Figure 2). We found that this compound exhibits insect feeding and ovipositional repellent activities against the yellow fever mosquito, Aedes aegypti (Culicidae) (25).

Table I. Antifeedant Activity of Azadirachtin and Two Derivatives in "Choice" Leaf Disk Assays Against *Spodoptera frugiperda* and *Leptinotarsa decemlineata*

Test insect	Test compound	Concentration (µg/cm²)	Effect
S. frugiperda	Azadirachtin	0.3	PC_{95}[a]
	22,23-Dihydroazadirachtin	0.3	PC_{95}[a]
	2',3',22,23-Tetrahydro-azadirachtin	0.3	PC_{95}[a]
L. decemlineata	Azadirachtin	60	PC_{50}[b]
	22,23-Dihydroazadirachtin	250	PC_{50}[b]
	2',3',22,23-Tetrahydro-azadirachtin	350	PC_{50}[b]

[a] PC_{95} is the minimum protective concentration of compound (µg/cm²) at which <5% of treated cotton leaf disks, while >95% of untreated leaf disks, are eaten in a "choice" assay.

[b] PC_{50} is the minimum protective concentration of compound (µg/cm²) at which <5% of treated potato leaf disks, while >50% of untreated leaf disks, are eaten in a "choice" assay.

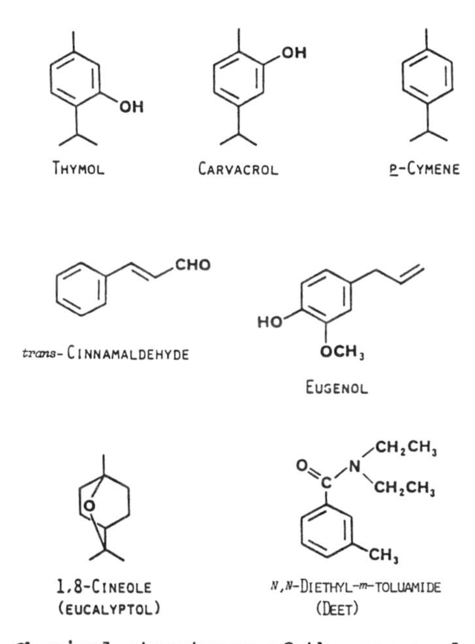

Figure 2. Chemical structures of the compounds tested for mosquito ovipositional repellent activity.

In the present study, we have expanded our investigation of biologically active plant volatile oil constituents to include several other naturally occurring compounds, namely eugenol, thymol, carvacrol, and trans-cinnamaldehyde, on the basis of availability and known biological activity (Figure 2). For example, eugenol is used commercially as an attractant for the Japanese beetle, Popillia japonica (Scarabaeidae) (2). Thymol and carvacrol are known antiseptic and anthelmintic agents and trans-cinnamaldehyde is used in flavoring (26). p-Cymene was included in this study because of its structural similarity to thymol and carvacrol, and Deet (N,N-diethyl-m-toluamide), a commercially available insect repellent, was included as a standard (Figure 2).

Gravid A. aegypti females were given a choice of ovipositing on paper toweling lining glass crystallizing dishes (70 x 50 mm) containing either tap water or one of the test compounds in tap water at a concentration ranging from 0.01-0.8% w/v (25). Eggs laid in the dishes were counted and results are given in Table II as the number of eggs laid in the treated dishes divided by the number of eggs laid in the control dishes, multiplied by 100 (T/C x 100%).

From Table II, it is apparent that, as A. aegypti ovipositional repellents, the relative activity of thymol>carvacrol = trans-cinnamaldehyde>eugenol>Deet>1,8-cineole>p-cymene. Even at the lowest concentration tested (0.01%), thymol, carvacrol, and trans-cinnamaldehyde were effective ovipositional repellents. Eugenol was less effective, yet was as active at 0.01% as Deet was at 0.2%. 1,8-Cineole was slightly less active than Deet. In general, the oxygenated aromatic compounds were more effective than the oxygenated cycloaliphatic compound, 1,8-cineole. p-Cymene, the parent hydrocarbon of thymol and carvacrol, was the weakest of the compounds tested. These results are in agreement with previous findings which suggest that the best insect repellents are oxygenated, somewhat polar compounds in the molecular weight range of 100-250 (27, 28).

Although several of the natural plant phenolics and terpenoids we tested exhibited significant activity as mosquito ovipositional repellents against gravid A. aegypti females, further studies involving other natural plant products, synthetic analogs, and combinations of these are needed to optimize any new insect repellent formulations. Insect repellents presently available commercially such as Deet may be deleterious to humans, causing central nervous system disturbances (26). Hopefully, repellents such as thymol (presently used commercially in mouthwash) may prove as effective, yet safer to use.

Insect Growth Inhibitors

In order to identify compounds which may be useful as sources and models of new insect control agents, we have been screening over 800 species of xerophytic plants, mostly from arid and semi-arid regions of the western U.S.A., for activity against insects. For example, we have isolated mosquito larvicidal chromenes from H. fitchii (24) and insect growth inhibitory ellagitannins and ellagic acid from sticky geranium, Geranium viscosissimum (Geraniaceae) (29).

Table II. Ovipositional Repellent Activity of Naturally Occurring Phenolics, 1,8-Cineole, and Deet Against Female *Aedes aegypti*

Compounds		(100) T/C[a] ± S.D.[b]			
Conc. (% w/v):	0.8%	0.4%	0.2%	0.1%	0.01%
Thymol	0	0	0	0	1.8±2.9
Carvacrol	0	0	0	1.0±1.1	6.2±8.6
trans-Cinnamaldehyde	0	0	0.1±0.1	0.4±0.5	2.7±1.8
Eugenol	0	0	0.3±0.6	6.4±3.6	30.4±15.8
Deet	0	5.8±2.5	28.9±35.2	100.7±110.3	94.8±53.1
1,8-Cineole	2.0±2.0	13.7±11.6	42.1±2.4	--	--
p-Cymene	12.7±6.5	--	--	--	--

[a] $(100)T/C = \frac{\text{Eggs oviposited in treated dishes}}{\text{Eggs oviposited in control dishes}} \times (100)$; a value of zero indicates that no eggs were laid in the treated dishes; in practice, a value of (100)T/C > 30 indicated weak (or ineffective) repellency.

[b] S.D. = Standard deviation; for each data point, n ≥ 3.

Although ca. 10% of the plant species we have screened exhibited activity against insects in their methanolic extracts, only turpentine broom, Thamnosma montana (Rutaceae), showed significant activity in its hexane extract. For example, the hexane extract of T. montana, when incorporated into an artificial diet (30, 31) and fed to first instar H. virescens, caused growth inhibition and, at higher concentrations (ca. 2500 ppm), death.

Previous phytochemical work with T. montana had led to the isolation of a number of alkaloids and coumarins (32-35). Some of these compounds were reported to be biologically active. For example, Bennett and Bonner (32) reported the isolation of furanocoumarins from T. montana as plant growth inhibitors. Our analysis of the insect growth inhibitory hexane extract by gas chromatography/mass spectrometry (GC/MS) revealed it to consist largely of linear furanocoumarins (Figure 3). These compounds are known to exert a number of biological effects, including dermal photosensitization and antibacterial, antifungal, molluscicidal, allelopathic, insecticidal, and insect antifeedant activities (36, 37).

We isolated quantities of several of these furanocoumarins sufficient for further structural analysis by ^{1}H-NMR, UV, and IR spectroscopies. In addition, quantities of four of these compounds, namely psoralen, xanthotoxin, bergapten, and isopimpinellin (Figure 3) were readily available commercially in quantities sufficient for bioassay.

We tested these four furanocoumarins individually in the same previously described artificial diet bioassay (28°C, 80% r.h., 16:8 L/D photoperiod) with which we initially detected growth inhibitory activity against H. virescens (30-31). Following 12 days of exposure to treated diet, we compared the larval weights to control larval weights. Potency was determined as the effective concentration (EC_{50}) of each furanocoumarin added to the diet necessary to cause a 50% reduction in weight. EC_{50} values were determined from log probit paper and the confidence limits were determined by the method of Litchfield and Wilcoxon (38).

We found that the four furanocoumarins tested were of similar potency in our bioassay (EC_{50} = 110-115 ppm) (Table III). We also found a similar potency (EC_{50} = 105-150 ppm) when the same bioassay was performed with the same compounds in complete darkness in lieu of a 16:8 L/D photoperiod (generated with a Cool White bulb) used initially to test the compounds (Table III). Psoralen is highly phototoxic, as are xanthotoxin (with a methoxy group at position 4) and bergapten (with a methoxy group at position 9), to certain bacteria, fungi, nematodes, insects, birds, and mammals (37, 39-41). These electrophilic compounds, due to their furanocoumarinic ring, form photodiadducts with nucleophilic pyrimidine bases in DNA when irradiated with UV or visible light (42). However, the introduction of methoxy groups at both positions 4 and 9, as in isopimpinellin, causes the compound to become less electrophilic, and hence, nonphototoxic (43). In our bioassay, all four compounds were equally potent in both the light and dark indicating that the insect growth inhibitory activity was not due to phototoxicity. Instead, the larvae ingested less treated than control diet and therefore gained less weight (Table IV). For example, 1000 ppm of

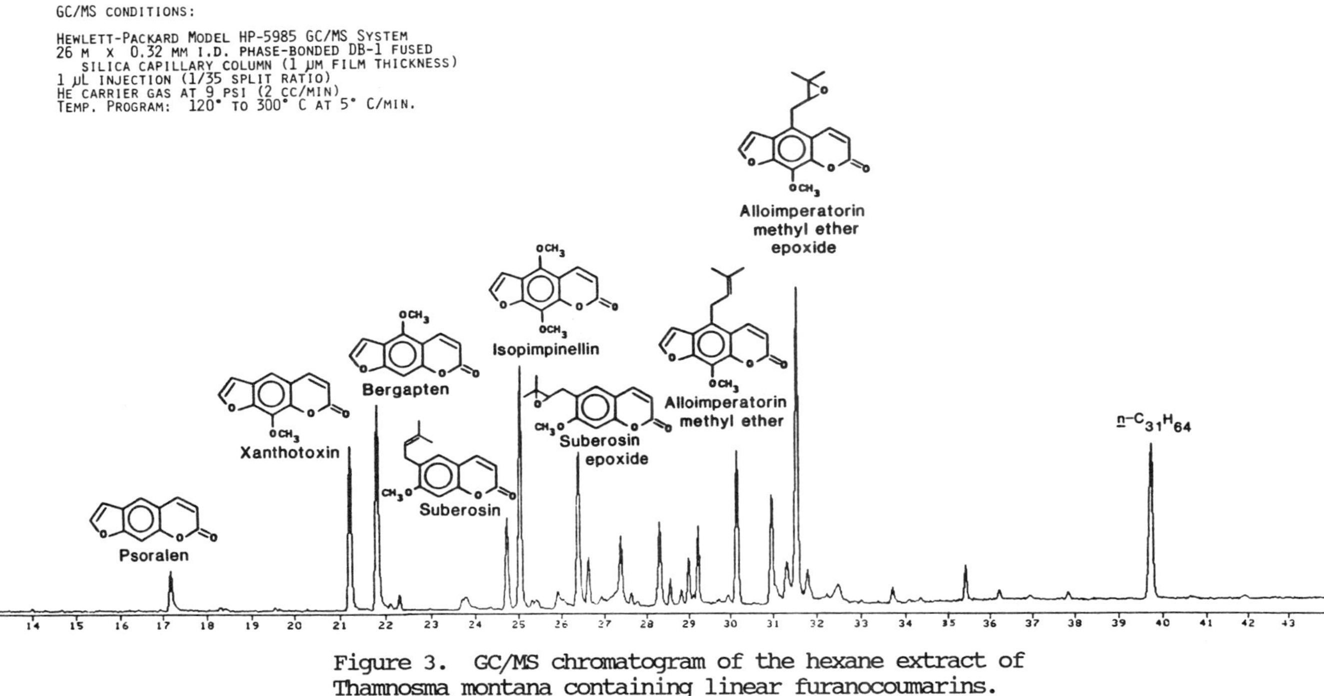

Figure 3. GC/MS chromatogram of the hexane extract of Thamnosma montana containing linear furanocoumarins.

Table III. Growth Inhibitory Activity of Four Linear Furanocoumarins from *Thamnosma montana* Fed in an Artificial Diet to First Instar Larvae of *Heliothis virescens*; Bioassays Conducted in a 16:8 L/D Photoperiod or in Complete Darkness for 12 Days

Test Compound	16:8 L/D Photoperiod		24-h Darkness	
	EC_{50}[a] (ppm)	Confidence Limits[b]	EC_{50} (ppm)	Confidence Limits
Psoralen	110	72-169	105	69-169
Xanthotoxin	110	83-146	115	91-146
Bergapten	115	84-158	150	129-175
Isopimpine-llin	110	74-164	110	72-168

[a] EC_{50} is the effective concentration of additive necessary to reduce larval growth to 50% of the control values.

[b] Confidence limits were determined using the method of Litchfield and Wilcoxon (38).

Table IV. Nutritional Indices of Fifth Instar Larvae of *Heliothis virescens* Fed Two Furanocoumarins from *Thamnosma montana* in an Artificial Diet; Bioassays Conducted in a 16:8 L/D Photoperiod for 3 Days

Test Compound (ppm)	Dry Wt. Gain (mg)	Dry Wt. Ingested (mg)	Approximate Digestibility	Efficiency of Conversion of Ingested Food
Control	35.7±9.9	261.0±67.6	45.0±7.4	14.3±4.4
Bergapten (500 ppm)	16.5±4.5*	163.2±63.5*	56.9±7.6	10.9±3.4
Bergapten (1000 ppm)	7.9±3.9*	69.8±42.5*	34.6±12.8	12.8±5.6
Isopimpinellin (1000 ppm)	7.0±4.6*	64.2±39.3*	52.9±18.9	12.6±9.8

*$p < 0.05$ using one way analysis of variance and multiple comparisons using Bonferroni's Inequality.

either the phototoxic bergapten or the nonphototoxic isopimpinellin fed to fifth instar H. virescens resulted in ca. 75% reduction in the dry weight of diet ingested (from 261.0 to 69.8 and 64.2 mg, respectively). This reduced ingestion in turn resulted in ca. 80% reduction in the dry body weight gained (from 35.7 to 7.9 and 7.0 mg, respectively). Bergapten fed at 500 ppm was also significantly active, causing a reduction in diet ingestion and body weight gain (Table IV). None of the treatments affected the approximate digestibility or the efficiency of conversion of ingested food of H. virescens (Table IV) (23,44).

Dark toxicity of furanocoumarins, as opposed to phototoxicity, has been observed by others. For example, Berenbaum (37) observed toxicity associated with nonphototoxic furanocoumarins in assays with the corn earworm, Heliothis zea (Noctuidae). Others have observed insect feeding repellent and antifeedant activities that are independent of ultraviolet light. For example, Yajima et al. (45) reported greater antifeedant activity against the tobacco cutworm, Spodoptera litura (Noctuidae), with isopimpinellin, a nonphototoxic furanocoumarin, than with xanthotoxin, a highly phototoxic compound.

Some attempts have been made to define the SAR of various furanocoumarins as insect antifeedants. For example, phloroglucinol-type furanocoumarins, such as bergapten and isopimpinellin, were found to be much more potent as antifeedants against S. litura than were pyrogallol-type, such as xanthotoxin, or resorcinol-type, such as psoralen, furanocoumarins (46). Bergapten and isopimpinellin were also found to be more potent than xanthotoxin against L. decemlineata, but not against the armyworm, Mythimna unipuncta (Noctuidae) (47). We found no difference in the antifeedant activity of bergapten, isopimpinellin, and xanthotoxin in preliminary leaf disk assays against S. frugiperda.

As suggested by the foregoing, there are various chemical approaches to the control of insect pests. Examples given include chemical effects on feeding with antifeedants (e.g., limonoids), on oviposition with repellents (e.g., phenolics), and on growth and development with growth inhibitors (e.g., furanocoumarins). Natural plant compounds can be expected to play an increasingly significant role in these various chemical approaches, both as models or leads for the chemical synthesis of structurally or topographically related mimics, and as sources of insect control agents, especially for local use and for the household insecticide market.

Acknowledgments

The authors thank Martina Janko and Lynette Nielson for technical assistance. This work was supported in part by a grant awarded by the U.S. National Science Foundation (PCM-8361004).

Literature Cited

1. Klocke, J.A. In "Allelochemicals: Role in Agriculture and Forestry"; Waller, G.R., Ed.; ACS Symposium Series No. 330, American Chemical Society: Washington, D.C., 1987; pp. 396-415.

2. Klocke, J.A. In "Economic and Medicinal Plant Research"; Wagner, H.; Hikino; H.; Farnsworth, N.R., Eds.; Academic Press: New York, 1989; Vol. 3, in press.
3. Kubo, I.; Nakanishi, K. In "Host Plant Resistance to Pests"; Hedin, P.A., Ed.; ACS Symposium Series No. 62, American Chemical Society: Washington, D.C., 1977; pp. 165-178.
4. Metcalf, R.L.; Luckmann, W.H. "Introduction to Insect Pest Management"; John Wiley & Sons: New York, 1982; 2nd ed.
5. Chapman, R.F. Bull. Entomol. Res. 1974, 64, 339-363.
6. Ascher, K.R.S.; Nissim, S. Int. Pest Control 1965, 7, 21-24.
7. Rice, E.L. "Pest Control with Nature's Chemicals: Allelochemics and Pheromones in Gardening and Agriculture"; University of Oklahoma Press: Norman, 1983; pp. 110-145.
8. Schoonhoven, L.M. Entomol. Exp. Appl. 1982, 31, 57-69.
9. Casida, J.E. In "Natural Products for Innovative Pest Management"; Whitehead, D.L.; Bowers, W.S., Eds.; Pergamon Press: New York, 1983; pp. 109-125.
10. Munakata, K. In "Chemical Control of Insect Behavior, Theory and Application"; Shorey, H.H.; McKelvey, J.J. Jr., Eds.; John Wiley & Sons: New York, 1977; pp. 93-102.
11. Jacobson, M.; Reed, D.K.; Crystal, M.M.; Moreno, D.S.; Soderstrom, E.L. Entomol. Exp. Appl. 1978, 24, 448-457.
12. Reed, D.K.; Jacobson, M.; Warthen, J.D. Jr.; Uebel, E.C.; Tromley, N.J.; Jurd, L.; Freeman, B. "Cucumber Beetle Antifeedants: Laboratory Screening of Natural Products"; Technical Bulletin Number 1641, U.S. Department of Agriculture - Science and Education Administration: Washington, D.C., 1981.
13. Ley, S.V. In "Recent Advances in the Chemistry of Insect Control"; Janes, N.F., Ed.; The Royal Society of Chemistry, Burlington House: London, 1985; pp. 307-322.
14. Nakanishi, K. In "Insect Biology in the Future"; Locke, M.; Smith, D.S., Eds.; Academic Press: New York, 1980; pp. 603-612.
15. Kubo, I.; Nakanishi, K. In "Advances in Pesticide Science"; Geissbuehler, H., Ed.; Pergamon Press: Oxford, 1979; Part 2, pp. 284-294.
16. Luteijn, J.M. Thesis, University of Wageningen, 1982.
17. Nakanishi, K. J. Nat. Prod. 1982, 45, 15-26.
18. Gill, J.S.; Lewis, C.T. Nature 1971, 232, 402-403.
19. Warthen, J.D. Jr. "Azadirachta indica: A Source of Insect Feeding Inhibitors and Growth Regulators"; Agricultural Reviews and Manuals ARM-NE-4, U.S. Department of Agriculture - Science and Education Administration: Beltsville, MD, 1979.
20. Yamasaki, R.B.; Klocke, J.A. J. Agric. Food Chem. 1987, 35, 467-471.
21. Pflieger, D.; Muckensturm, B. Tetrahedron Lett. 1987, 28, 1519-1522.
22. Ley, S.V.; Santafianos, D.; Blaney, W.M.; Simmonds, M.S.J. Tetrahedron Lett. 1987, 28, 221-224.
23. Barnby, M.A.; Klocke, J.A. J. Insect Physiol. 1987, 33, 69-75.
24. Klocke, J.A.; Balandrin, M.F.; Adams, R.P.; Kingsford, E. J. Chem. Ecol. 1985, 11, 701-712.
25. Klocke, J.A.; Darlington, M.V.; Balandrin, M.F. J. Chem. Ecol. 1987, 13, 2131-2141.

26. Windholz, M.; Budavari, S.; Blumetti, R.F.; Otterbein, E.S. "The Merck Index"; Merck and Co.: Rahway, New Jersey, 1983; 10th ed., pp. 261-262, 326, 412, 1347.
27. Wright, R.H. Scient. Am. 1975, 233, 104-111.
28. Dethier, V.G. Annu. Rev. Entomol. 1956, 1, 181-202.
29. Klocke, J.A.; Van Wagenen, B.; Balandrin, M.F. Phytochemistry 1986, 25, 85-91.
30. Chan, B.G.; Waiss, A.C. Jr.; Stanley, W.L.; Goodban, A.E. J. Econ. Entomol. 1978, 71, 366-368.
31. Kubo, I.; Klocke, J.A. In "Plant Resistance to Insects"; Hedin, P.A., Ed.; ACS Symposium Series 208, American Chemical Society: Washington, D.C., 1983; pp. 329-346.
32. Bennett, E.L.; Bonner, J. Am. J. Bot. 1953, 40, 29-33.
33. Dreyer, D.L. Tetrahedron 1966, 22, 2923-2927.
34. Kutney, J.P.; Verma, A.K.; Young, R.N. Tetrahedron 1972, 28, 5091-5104.
35. Chang, P.T.O.; Cordell, G.A.; Aynilian, G.H.; Fong, H.H.S.; Farnsworth, N.R. Lloydia 1976, 39, 134-140.
36. Scott, B.R.; Pathak, M.A.; Mohn, G.R. Mutat. Res. 1976, 39, 29-74.
37. Berenbaum, M. Rec. Adv. Phytochem. 1985, 19, 139-169.
38. Litchfield, J.T. Jr.; Wilcoxon, F. J. Pharmacol. Exp. Ther. 1949, 96, 99-113.
39. Ivie, G.W. In "Effects of Poisonous Plants on Livestock"; Keeler, R.F.; Van Kampen, K.R.; James, L.F., Eds.; Academic Press: New York, 1978; pp. 475-485.
40. Song, P.-S., Tapley, K.J. Jr. Photochem. Photobiol. 1979, 29, 1177-1197.
41. Berenbaum, M. Evolution 1983, 37, 163-179.
42. Kanne, D.; Straub, K.; Hearst; J.E.; Rapoport, H. J. Am. Chem. Soc. 1982, 104, 6754-6764.
43. Musajo, L.; Rodighiero, G. Experientia 1962, 18, 153-200.
44. Reese, J.C.; Beck, S.D. Ann. Entomol. Soc. Am. 1976, 69, 59-67.
45. Yajima, T.; Kato, N.; Munakata, K. Agric. Biol. Chem. 1977, 41, 1263-1268.
46. Yajima, T.; Munakata, K. Agric. Biol. Chem. 1979, 43, 1701-1706.
47. Muckensturm, B.; Duplay, D.; Robert, P.C.; Simonis, M.T.; Kienlen, J.-C. Biochem. Syst. Ecol. 1981, 9, 289-292.

RECEIVED November 2, 1988

Chapter 11

Azadirachtins

Their Structure and Mode of Action

Heinz Rembold

Max Planck Institute for Biochemistry, Insect Biochemistry, D–8033, Martinsried, Federal Republic of Germany

Seven tetranortriterpenoids were isolated from neem (Azadirachta indica A. Juss) seed by use of the Epilachna varivestis bioassay. They were structurally elucidated and two of them chemically modified. All these compounds are similar to the main compound, azadirachtin A, as well in quality and quantity of their biological activity as in their chemical structure. Based on these data, a reduced chemical structure is proposed. - Azadirachtins are insect growth inhibitors. They interfere with the neuroendocrine regulation of juvenile and molting hormone titers. Main cellular targets are the Malpighian tubules and the corpus cardiacum (CC) of the insect. - In the CC the azadirachtins reduce the turnover of neurosecretory material, as demonstrated by poor labelling with ^{35}S-cysteine. Consequently, levels of the morphogenetic juvenile and molting hormones are shifted and concomitantly decreased after azadirachtin injection. In this way, metamorphosis of the juvenile insect is inhibited and reproduction of the adult as well.

The high biological activity of azadirachtin, a tetranortriterpenoid which was isolated from seed kernels of the neem tree, Azadirachta indica, is well established. Though a strong antifeedant to locusts (1) and to other insects of several taxa, it also acts as a potent growth inhibitor at microgram levels (2-4). The most defined effects are (a) delay and/or inhibition of molt into the successive instar, (b) disturbance of the molting pro-

0097–6156/89/0387–0150$06.00/0

cess, and (c) delay, disturbance or inhibition of ovarian development.

Several studies were made during recent years to elucidate the modifications in the endocrine control mechanisms induced by azadirachtin, that led to the observed growth inhibiting effects. The salient feature that emerges out of such studies on the migratory locust, Locusta migratoria, is the ultimate change (reduction and delay) in the titer of the morphogenetic hormones, namely ecdysone (2,3) and juvenile hormone (5,6). However, it is not clear so far whether such an effect on hormone titers is a direct or an indirect one.

A direct effect of azadirachtin on ecdysone synthesis by the prothoracic glands has been ruled out in a study on Bombyx mori (7). Evidence available so far suggests that azadirachtin may block the release of several trophic factors located in the central nervous system (3). The chemical identity of these factors, like the prothoracicotropic and allatotropic hormones from locust, is still awaited. Hence comparison of the stainability of neurosecretory material in histological preparations with paraldehyde fuchsin has been widely followed to express the neurosecretory activity. It was suggested (5) that in L. migratoria, azadirachtin treatment leads to accumulation of stainable neurosecretory material in the corpus cardiacum, the neurohemal organ with storage and release function.

Such natural or synthetic insect growth inhibitors are of interest for the chemist in his search for new chemical structures aiming more selectively at such targets which are different from those of the present broad-spectrum neurotoxic insecticides. One interesting target of such growth inhibiting compounds is the insect's specific hormone system. This could be disturbed by a growth inhibitor either in a direct way through the application of hormone analogs like the juvenoids, or in a more indirect way through an interference with the neuroendocrine regulation of its peripheric hormones, the juvenile and the molting hormones. Other alternatives are the antifeedants. If they can induce starvation in the herbivore, they would indirectly cause developmental deviants (10). However, usually antifeedants drive the herbivorous insect larva to the untreated growing parts of the plant where they can recover if the plant is not treated again with feeding deterrent. Similar arguments also fairly often speak against the use of repellents for plant protection purposes.

Finally, studies on new types of insecticides are important for controlling such insects which are vectors of diseases. It may be possible, that the host-specific parasite can be controlled by insect growth inhibitors like the azadirachtins by disturbing the endocrine and consequently physiological situation of the insect host. With such an approach in mind it may even be possible to cure the insect vector from its parasite.

The Epilachna Bioassay

A bioassay for detection of a whole group of natural insect growth inhibitors, as present in neem (8), has to combine high sensitivity for growth disrupting compounds with high tolerance for antifeedants. The Mexican bean beetle, *Epilachna varivestis*, combines these two attributes under simple test conditions. Two tests have been described for routine assays, a Petri dish test for individual, and a cage test for groups of larvae (8). One advantage of this bioassay is, that the test insects are easily reared on whole plants or leaves of the bean, *Phaseolus vulgaris*, which can be grown in the greenhouse year round.

In the *Petri dish test*, one bean leaf is put into each of 20 plastic Petri dish covers (9,2 cm diam.) which already contain a moist filter paper. The leaf is then inversely covered with the same size Petri dish bottom which has a hole of 4 cm diameter and presents a defined leaf surface area. The test material, dissolved in 0,2 ml methanol, is equally distributed on the exposed 12,6 cm^2 of leaf area. Then the dish is covered with another Petri dish top. A weighed, freshly molted fourth instar larva is released into each dish. After 24 and 48 hours, the weight gains of all larvae are calculated. After the two days on the treated, they are transferred to untreated bean leaves where their further development is followed.

For routine estimations, the *cage test* needs less expenditure of work and also eliminates a possible fumigant effect of volatile by-products from the chromatographic fractions to be tested. An aluminium-framed cage (30 x 25 x 32 cm) with a wooden bottom, a glass top and the sides covered with a mesh-net for ventilation, can be used. Ten young bean plants, with twenty primary leaves altogether, are uniformly sprayed with 20 ml methanolic solution and placed inside the cage after drying. Twenty freshly molted fourth instar larvae are released in each cage and new grown leaves are plucked off routinely. The treated bean plants are removed after 48 hours and replaced by untreated ones.

In both Petri dish and cage tests, two control treatments are run. In one control, the leaves are treated with the same volume of pure methanol as in the test group, and in the other plants are left untreated. The undisturbed development of the control groups from beginning of the fourth instar to the newly emerged adults takes eight days whereas after azadirachtin treatment the last insects may die after three weeks only. The tests have to be repeated if the concentration of the applied compound does not result in about 50% survival. On this basis, the MC_{50} (50% metamorphosis inhibiting concentration) as well as the LC_{50} (lethal concentration) values can be calculated.

By following the growth inhibiting effect with the *Epilachna* bioassay, Schmutterer and Rembold (9) isolated four growth inhibiting compounds from neem seed which did not affect larval feeding at concentrations which induced severe metamorphic disturbances. The most prominent growth inhibitor came out to be identical with azadirachtin which was already known to have, besides its more or less deterring also growth inhibiting activity in most if not all of the economically important insect orders. Also the other growth inhibitors from neem do not only inhibit insect growth in the *Epilachna* assay, but also that of other insects, and again at such concentrations which do not cause feeding inhibition (*8*).

The predominant compound which is eluted in the azadirachtin peak from silica gel makes up about 85% of the total growth inhibiting activity. Therefore it has been named azadirachtin A (*5*). Its former structure as proposed by Zanno *et al.* (*11*), has recently been reassigned by three laboratories (*12*-*14*) and now unequivocally gives the basis for a structural elucidation of the other azadirachtins by nmr spectroscopy. Azadirachtin A (I) is a highly oxidized tetranortriterpenoid with rings A and B *trans* connected, an epoxide ring at position 13, 14 and a tigloyl side chain at position 1. Three hydroxyl groups, at positions 7, 11, and 20, are free in the azadirachtin A molecule.

Azadirachtin B (II) was isolated by following its insect growth inhibiting effect in the *Epilachna* bioassay (*15*,*16*). It is with about 15% abundance the second most prominent azadirachtin in neem kernels. The tigloyl side chain is in position 3, in contrast to the location in azadirachtin A where it is located in position 1. This isomer also has a free hydroxyl group at position 1 and the hydroxyl group in position 11 is reduced to the deoxy compound. Altogether, both these isomeric azadirachtins carry three free hydroxyl groups.

The minor group of azadirachtins C - G (Figure 1) was also isolated by use of the *Epilachna* bioassay. These isomers are present in only minute amounts of one percent altogether in the crude azadirachtin mixture (*17*). As to be expected from their biological activity, they structurally look very much like the azadirachtins A and B. Substitution of the hydroxyl groups 1 (tigloyl) and 3 (acetyl) like in azadirachtin A is characteristic for azadirachtin C (VII, Table I), for which only a partial structure can be given out of nmr data, for azadirachtin D (III) with its ester group in position 4 reduced to methyl, and for azadirachtin E (IV) which is the naturally occurring 1-detigloyl azadirachtin A. Natural isomers of azadirachtin B with free 1-hydroxyl and 3-tigloyl substitution are azadirachtin F (V) with the ether bridge in position 19 reduced and opened by formation of a C-19 methyl, and azadirachtin G (VI) with a double bond formed instead of the 13,14-epoxide ring and with a hydroxyl group in position 17.

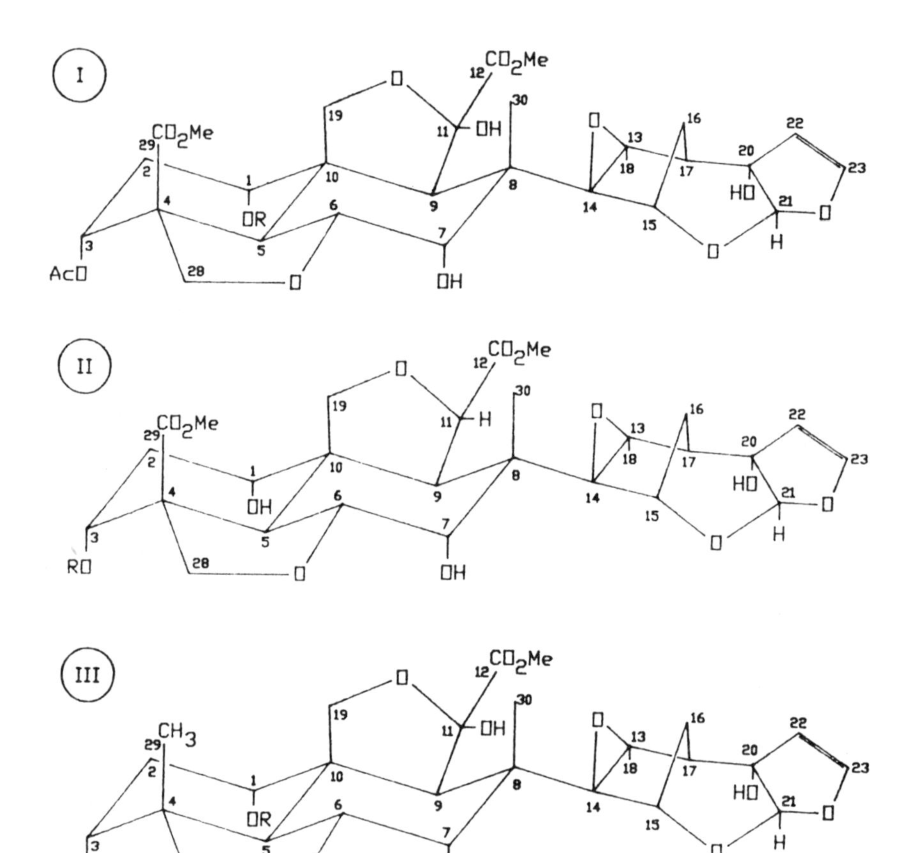

Figure 1a. Structures of the natural azadirachtins that have been isolated from neem seeds (<u>15</u>, <u>17</u>). R: tigloyl.

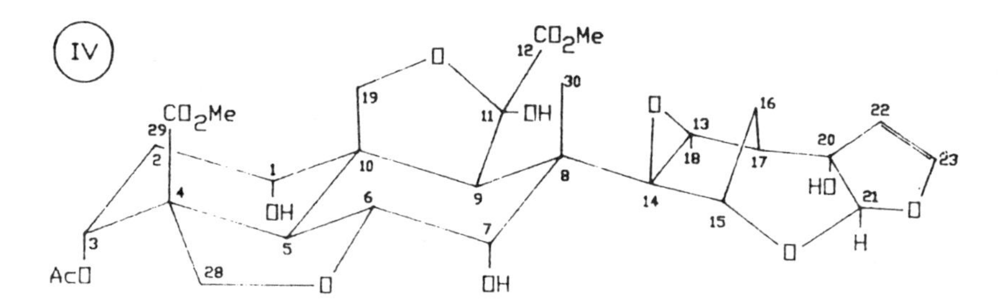

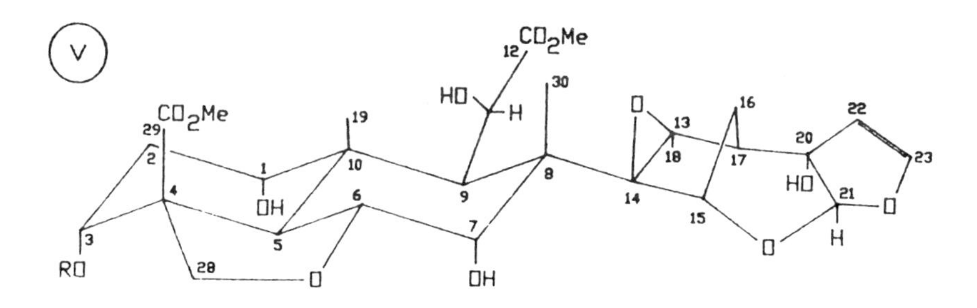

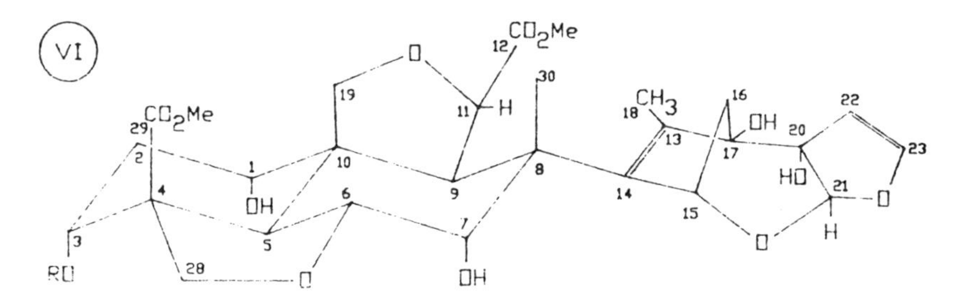

Figure 1b. Structures of the natural azadirachtins that have been isolated from neem seeds (15, 17). R: tigloyl.

The azadirachtin molecule is chemically labile and structural modifications of the natural product usually come out in low yields. Only hydrogenation of the 22,23-double bond, and correspondingly its tritiation (5), can be achieved in good yields. Altogether five chemical modifications of azadirachtin A and four of azadirachtin B have been described (15). They are collated in Table I, together with data on their biological activity and including the other natural azadirachtins, which now allows some conclusions on structure - activity relationships.

All the azadirachtins tested so far are under standard *Epilachna* test conditions of acute toxicity in concentrations above 1000 ppm. The larvae die within a few hours already. Pharmacological effects like on gut motility or diuresis are found at lower doses and have sometimes been falsely interpreted as showing the physiological mode of azadirachtin action already (18). In this pharmacological range of 10 - 100 ppm concentrations the azadirachtins act as phagodeterrents cum growth inhibitors. However, in their physiological range of 1 - 5 ppm concentrations they do not deter the *Epilachna* larva from feeding but act as potent growth inhibitors. The data of Table I are based on this physiological, exclusively growth disrupting effect of the azadirachtins.

Table I. Growth inhibition, expressed as lethal concentration for 50% of the test larvae (LC_{50}), in the *Epilachna* cage test (15,17,25)

#	Compound	LC_{50} (ppm)
I	azadirachtin A	1.66
IA	11-acetyl-azadirachtin A	8.68
IB	3-deacetyl-azadirachtin A	0.38
IC	22,23-dihydro-azadirachtin A	1.26
ID	23 -ethoxy-22,23-dihydro-azadirachtin A	0.74
IE	23 -ethoxy-22,23-dihydro-azadirachtin A	0.52
II	azadirachtin B	1.30
IIA	3-detigloyl-azadirachtin B	0.08
IIB	22,23-dihydro-azadirachtin B	0.28
IIC	3-detigloyl-3-(2-methylbutyryl)-22,23-dihydro-azadirachtin B	0.45
IID	23 -ethoxy-22,23-dihydro-13,14-deepoxy-17-hydroxy-azadirachtin B	>100
VII	azadirachtin C	12.97
III	azadirachtin D	1.57
IV	azadirachtin E	2.80
V	azadirachtin F	1.15
VI	azadirachtin G	7.69
	salannin	>100

The type of decalin ring substitution seems to be of high importance, if the 16 azadirachtin structures of Table I are compared. The highest biological activity is with both hydroxyl groups free (IIA). Also the ecdysteroid molecule has two unsubstituted hydroxyl groups in its A ring. However, the fundamental structural difference between these two bioactive molecules is that rings A and B are trans-connected in the azadirachtins and cis in the ecdysteroids. Another interesting position is the 22,23-double bond which is present in all the natural azadirachtins. Hydrogenation (IC, IIB) or addition of alcohol (ID, E) even increases biological activity, however. Azadirachtins B (II) and F (V) are significantly more active than azadirachtin A (I) and mainly its 11-acetyl derivative (IA). The most critical structural element seems to be the 13,14-epoxy group. Both IID and salannin (19) which both are without this epoxy group, are completely inactive as insect growth inhibitors. Also VI and VII are significantly less active. It seems that a ketal function at position C-21 and a free hydroxyl group at C-7 supports growth inhibition, although there are some facts which do not become clear from the present experimental data. With all these relationships in mind, a reduced structure (VIII, Figure 2) has been proposed to be primarily responsible for biological activity of the azadirachtin group (17). Substitution of the decalin ring by hydroxyl groups at C-1 and C-3 and a 13,14-epoxy group in correct steric distance from these two hydroxyl groups seems to be of primary importance for the biological effect. Minor effects are to be expected from a 7-hydroxy and a 21-ketal function. Also a reduced dihydrofuran ring or a side chain in correct distance from the epoxide group may add to the biological activity.

Effect on hormone titers

Azadirachtin strongly interferes with larval growth and development of all the insects which have been studied so far. The morphological effects are growth retardation, molting inhibition, or induction of malformations. In the adult insect, it inhibits egg maturation. Azadirachtin may also induce sterilization in several insect species (2,20, 21). Such effects can also be induced by hormone application and one may therefore wonder if azadirachtin mimics a hormone, eventually also one of the hormone metabolites.

Three main questions may help for a better understanding of its mode of action on the hormonal level. (i) How is the resorption/excretion of this drug in the insect? Does it act at a physiological or at a pharmacological concentration? This question was followed by measuring the excretion kinetics of the tritium labeled compound and will be discussed in some more detail. (ii) Is azadirachtin only antagonizing the ecdysteroid function or does it also affect the juvenile hormone titer, concomitantly or independent from the ecdysteroid titer? This question was followed in individual living insects. And finally (iii) does the drug accelerate or reduce protein turnover due to high affinity binding in the neurosecretory system?

[22,23-$^3H_2$]Dihydroazadirachtin A, which after Table I has the same physiological activity as azadirachtin A, was used to follow its excretion kinetics and organ specific incorporation in L. migratoria. Even if injected at such a low concentration as 0.1 µg/g, more than 50% of the drug is excreted unchanged within 24 hours (5,6). No difference is found in the pattern of azadirachtin excretion if adults injected 1, or 2 days after emergence and last instar nymphs injected at a low and high dose are compared. Irrespective of the dose injected (0.41 - 2.47 µg/g), between 68 and 82% of the total measured radioactivity is excreted, most of it as unchanged ditritioazadirachtin, within the first 24 hours after application. The material which is retained in the treated locust can not be mobilised by injection of unlabeled dihydroazadirachtin. It is therefore tightly bound, most of it in the Malpighian tubules, followed by the gut and ovaries (6). Even 15 days after injection, the intact labeled material could be isolated from the organs. In absolute terms, it was contained in the Malpighian tubules in a concentration of about 25 ng/g, and in the gut and ovaries in about 3 ng/g, whereas the remaining biomass contained less than 1 ng/g (Rembold, H.; Müller, Th.; Subrahmanyam, B. Z. Naturforsch., in press.). Consequently, the physiological concentration, as discernible by the retained amount of azadirachtin after injection of 2.5 µg/g, was about 1 percent in the Malpighian tubules and even only 1 per mille in the gut and ovary. These extremely low concentrations speak in favor of a high affinity binding of the drug and against a pharmacological or toxic effect already.

Before discussing the results obtained from the histological autoradiography of the corpus cardiacum, the question must be answered, in what way azadirachtins affect the titers of juvenile and molting hormone. Clear results could be obtained, when the hemolymph titer of the two hormones was followed in individual animals during their gonotrophic cycle. Precondition for such a study is the existence of a highly selective and quantitative hormone assay which is reliable at low quantities already. Gene activity during metamorphosis and reproduction is controlled by hormone/receptor titers. Obviously, affinity of the receptor determines the amount of hormone necessary for releasing a signal. The dissociation constant (K_D) values for the receptor-hormone complexes are estimated to be in the range of 10^{-8} and 10^{-11} M. These are concentrations of few picograms per 0.01 ml of insect hemolymph which must be taken from an individual locust per 24 hrs.

Whereas for quantitative ecdysterone determinations a combination of high-performance liquid chromatography in combination with radioimmuno assay can be used, this is impossible for the minute amounts of juvenile hormone due to cross-reactivity of the immunoassay with other lipids. A physicochemical method which combines microderivatisation of the juvenile hormone to its 10-dimethyl(nonafluorohexyl)silyloxy-11-methoxy derivative which after simple purification steps is finally separated and quantified through capillary GLC combined selected ion monitoring mass spectrometry (22). The enormous advantage of this chemical technique is the use of an internal standard which provides quantitative results for each of the four juvenile hormones. By following both molting and juvenile hormone titers in each hemolymph sample during a whole gonotrophic cycle, and in comparison with the untreated control, a synchronous endocrine control has become clear. Both juvenile hormone and ecdysone titers are affected after a single injection of azadirachtin. Their peaks are identically shifted to a longer time period, mostly without decrease in their intensity. A similar retardation, and decrease, comes out for the hemolymph proteins and primarily for vitellogenin (6).

Inhibition of neurosecretion turnover

It has become clear, primarily from studies with *L. migratoria*, that the salient feature of azadirachtin function is the ultimate change (reduction and delay) in the titer of the morphogenetic hormones. However, it is not yet clear how to understand the obvious blocking of the release of trophic factors from the neuroendocrine system. There are two possibilities for explaining the obvious accumulation of PAF stainable neurosecretory material (NSM) in the neurohemal organ and neurosecretory cells of the azadirachtin treated locusts. (i) Interference of the drug with the feedback control of hormone titers and consequently an increase of NSM synthesis and decrease of its release. (ii) The other explanation could be the concomitant blocking of release and of NSM synthesis; a sharp drop in precursor incorporation would be the consequence. The latter could be demonstrated (Subrahmanyam, B.; Müller, Th.; Rembold, H. *J. Insect Physiol.*, in press) by following incorporation of 35-S-cysteine into the CC proteins. The method makes use of the fact, that neurosecretory proteins are rich in the sulfur containing amino acids, cysteine and cystine, which are incorporated into newly synthesised neurosecretory protein primarily.

The corpora cardiaca from azadirachtin treated *L. migratoria* females show a very poor NSM turnover. Immediately after injection of the labeled amino acid, there is a raise in radioactivity which is much higher in the CC of the azadirachtin treated insects. Turnover of free cysteine is obviously inhibited by azadirachtin. There is no qualitative change in the ^{35}S-labeled proteins of this organ, as shown by electrophoresis and autoradiography of the gels. However, in the azadirachtin treated insects the transport of labeled protein from the brain to the corpus cardiacum and its release are at a very low level, though not completely inhibited. Hence, disturbance or inhibition of ovarian development in the azadirachtin treated females is mainly due to the changes induced in the endocrine events by the poor turnover of neurosecretory proteins. A histological study of brain and CC, which will be discussed in the following, shows a significant accumulation of NSM in the secretory reservoir. The increased stainability can therefore be interpreted as only to be due to accumulation of NSM over a longer period of time together with slow release, and does not denote an increase in synthesis.

Organ-specific incorporation of azadirachtin

Autoradiographs of the organ with highest specific radioactivity, the Malpighian tubules, from insects that received a high specific activity of labeled dihydroazadirachtin A show an intense accumulation of silver grains all along the basal region of the tubule, and the cytoplasm around the nucleus. The intensly labeled tubules form a radioactive trace. The apical region (microvilli), which is limiting the lumen of the tubule, is relatively unlabeled as well as the nucleus where only its membrane shows high labeling. The autoradiographic study thus provides evidence for the accumulation of dihydroazadirachtin A in two histological regions of the Malpighian tubules namely, the basal region and the cytoplasm around the nucleus (Rembold, H.; Müller, Th.; Subrahmanyam, B. _Z. Naturforsch._, in press.). This clear localisation of chemically unchanged dihydroazadirachtin in the basal and inner regions of the tubule even after a long time period suggests its organ and region specific concentration to high-affinity binding sites.

The autoradiographic study of the retrocerebral complex of _L. migratoria_ brings out that azadirachtin A has a free access to the corpus cardiacum rather than into the brain (Subrahmanyam, B.; Müller, Th.; Rembold, H. _J. Insect Physiol._, in press.). The labeled compound does not penetrate the brain but is located only at the peripheral blood brain - barrier, the perineurium - neuroglia complex. In the corpus cardiacum, the axons that terminate in the storage lobe are intensely labeled. The cytoplasm of the intrinsic secretory cells of the glandular lobe is also intensely labeled whereas the nuclei of these cells remain unlabeled (Fig. 3). The corpus cardiacum acts as a neurohemal organ which releases trophic peptide factors into the hemolymph. The neurosecretory cells of the brain release their products from the axon endings in the corpus cardiacum. For the

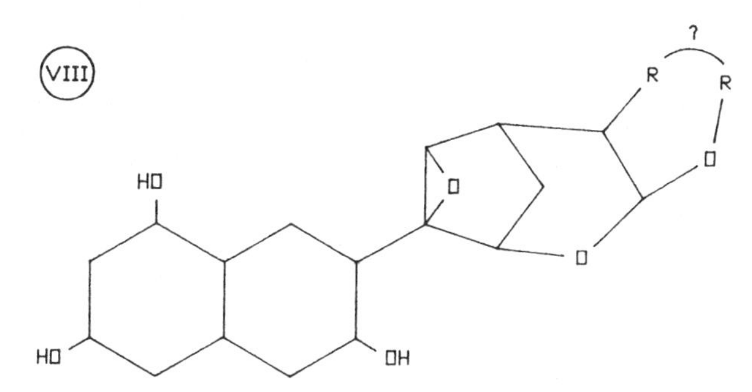

Figure 2. Proposal for a reduced bioactive azadirachtin structure (17).

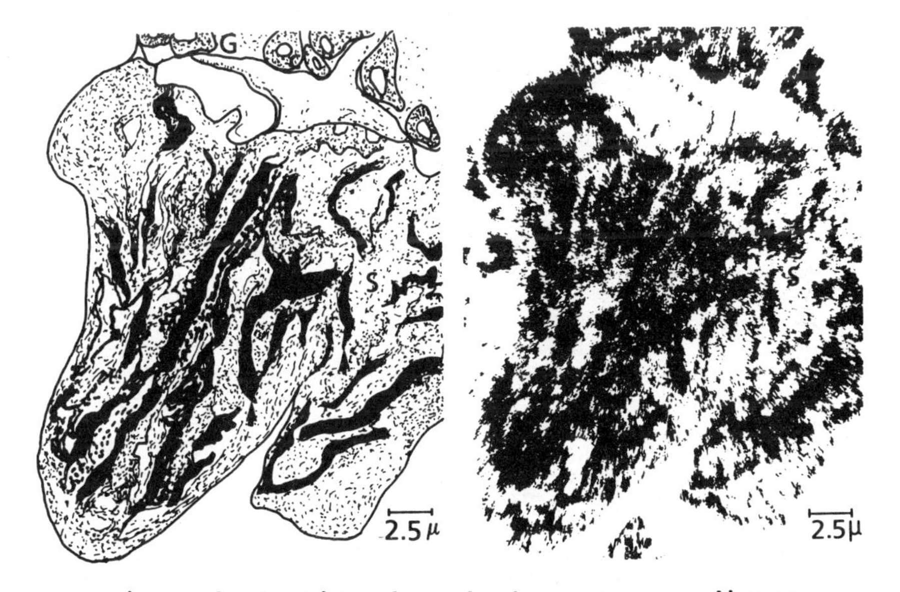

Figure 3. Section through the corpus cardiacum of *L. migratoria* from a female adult, 5 days after injection of 3 μg [22,23-$^3H_2$]dihydroazadirachtin A, showing the intense distribution of silver grains (right) developed on the autoradiogram. For better information the same view is drawn on the left side. S: Storage lobe, G: part of glandular lobe, C: cytoplasm of secretory cells. The arrows are indicating the heavily labeled axons.

efficient release of neurohormones into the hemolymph, the axon terminals are extensively branched and are devoid of the glial ensheathment which protects the brain, so that their endings are totally exposed to the external environment. Hence, the axons terminating in the CC are heavily labeled with azadirachtin which possibly blocks the release of neurohormones into the hemolymph. The low synthesis of neurosecretory proteins as well as their accumulation in the retrocerebral complex are then the result of a feed back regulation operating between the release and synthesis of neurosecretory material.

Azadirachtins, promising botanical pesticides

Three quarters or at least eight hundred thousand of all animal species are members of the insect class. Only few of them are agricultural pests or vectors of harmful diseases. One estimates a world-wide economic loss, caused by insects, of more than 20 mrd USD in agriculture although highly effective insecticides are applied. However, the problems associated with conventional broad-spectrum insecticides and a rapidly increasing number of insecticide tolerant or even resistant pests are all too apparent.

Search for more selective and biodegradable insecticides, studies on host-plant resistance, insect attractants, use of natural enemies, autocidal techniques are some promising fields within integrated pest management strategies. However, progress can only be expected from new fields which must be endeavored by basic studies. Naturally occurring insecticides, also from weeds, must be found by use of easy and reliable bioassays. Such studies must also include investigations on the basis of host- parasite interactions in the case of many diseases, like Malaria, yellow fever, or Chagas disease, which are transmitted by insects. The group of azadirachtins seems to be an exceptionally promising model in the search for botanical pesticides. Their target is the insect-specific hormonal control of growth and development. As far as we know till now, they have no mammalian toxicity (Jacobson, M., personal communication), act at extremely low concentrations already due to their target specific binding to the corpus cardiacum and the Malpighian tubules, and they are easily biodegradable. The proposed reduced chemical structure now has to be synthesised and modified. The molecular mode of azadirachtin action still has to be elucidated. Knowing more about its target-directed mechanism may then help in tailoring a new generation of synthetic insect growth inhibitors.

Literature Cited

1. Butterworth, J. H.; Morgan, E. D. J. Insect Physiol. 1971, 17, 974-77.
2. Rembold, H.; Sieber, K.-P. Z. Naturforsch. 1981, 36c, 466-69.
3. Sieber, K.-P.; Rembold, H. J. Insect Physiol. 1983, 29, 523-27.
4. Mordue, A. J.; Evans, K. A.; Charlet, M. Comp. Biochem. Physiol. 1986, 85C, 297-301.
5. Rembold, H.; Forster, H.; Czoppelt, Ch.; Rao, P. J.; Sieber, K.-P. Proc. 2nd Int. Neem Conf. Rauischholzhausen; Schmutterer, H. and Ascher, K. R. S., Eds.; GTZ, Eschborn, 1983, pp 153-61.
6. Rembold, H.; Uhl, M.; Müller, Th. Proc. 3rd Int. Neem Conf. Nairobi; Schmutterer, H. and Ascher, K. R. S., Eds.; GTZ Eschborn, 1986, pp 289-98.
7. Koul, O.; Amanai, K.; Ohtaki, T. J. Insect Physiol. 1987, 33, 103-08.
8. Rembold, H.; Sharma, G. K.; Czoppelt, Ch.; Schmutterer, H. Z. Pflkrankh. Pflschutz. 1980, 87, 290-97.
9. Schmutterer, H.; Rembold, H. Z. angew. Entomol. 1980, 89, 179-88.
10. Slama, K. Acta Ent. Bohemoslov. 1978, 75, 65-82.
11. Zanno, P. R.; Miura, I.; Nakanishi, K.; Elder, D. L. J. Amer. Chem. Soc. 1975, 97, 1975-77.
12. Turner, C. J.; Tempesta, M. S.; Taylor, R. B.; Zagorski, M. G.; Termini, J. S.; Schroeder, D. R.; Nakanishi, K. Tetrahedron 1987, 43, 2789-2803.
13. Bilton, J. N.; Broughton, H. B.; Jones, P. S.; Ley, S. V.; Lidert, Z.; Morgan, E. D.; Rzepa, H. S.; Sheppard, R. N.; Slawin, A. M. Z.; Williams, D. J. Tetrahedron 1987, 43, 2805-15.
14. Kraus, W.; Bokel, M.; Bruhn, A.; Cramer, R.; Klaiber, I.; Klenk, A.; Nagl, G.; Pöhnl, H.; Sadlo, H.; Vogler, B. Tetrahedron 1987, 43, 2817-30.
15. Rembold, H.; Forster, H.; Czoppelt, Ch. Proc. 3rd Int. Neem Conf. Nairobi; Schmutterer, H. and Ascher, K. R. S., Eds.; GTZ Eschborn, 1986, pp 149-160.
16. Rembold, H.; Forster, H.; Sonnenbichler, J. Z. Naturforsch. 1987, 42c, 4-6.
17. Forster, H. Dr. Thesis, University of Munich, Munich, 1987.
18. Mordue, A. J.; Cottee, P. K.; Evans, K. A. Physiol. Entomol. 1985, 10, 431-37.
19. Henderson, R.; McCrindle, R.; Melera, A.; Overton, K. A. Tetrahedron 1968, 24, 1525
20. Koul, O. Z. angew. Entomol. 1984, 98, 221-223.
21. Rembold, H. Advances in Invertebrate Reproduction 3; Engels, W., Ed.; Elsevier, Amsterdam, 1984, pp 481-491.
22. Rembold, H.; Lackner, B. J. Chromatog. 1985, 323, 355-61.

RECEIVED November 2, 1988

Chapter 12

Naturally Occurring and Synthetic Thiophenes as Photoactivated Insecticides

J. T. Arnason[1], B. J. R. Philogène[1], Peter Morand[2], K. Imrie[1], S. Iyengar[1], F. Duval[1], C. Soucy-Breau[2], J. C. Scaiano[3], N. H. Werstiuk[4], B. Hasspieler[5], and A. E. R. Downe[5]

[1]Department of Biology, University of Ottawa, Ottawa, Ontario K1N 6N5, Canada
[2]Department of Chemistry, University of Ottawa, Ottawa, Ontario K1N 6N5, Canada
[3]Division of Chemistry, National Research Council, Ottawa, Ontario K1A 0R6, Canada
[4]Department of Chemistry, McMaster University, Hamilton, Ontario L8S 4M1, Canada
[5]Department of Biology, Queens University, Kingston, Ontario K7L 3N6, Canada

The secondary compound alpha-terthienyl derived from the plant family Asteraceae and related molecules are under consideration as a new class of photoactivated insecticides. Trials under tropical conditions indicate a very high level of activity as a larvicide for the malaria mosquito, _Anopheles gambiae_. There is no cross resistance to this compound in malathion resistant mosquito larvae. However tolerance observed in some insects can be related to metabolism and elimination of labelled compound. Over thirty synthetic analogues and derivatives have been produced to examine structure-activity relationships and singlet O_2 generating potential of the molecules.

At a recent ACS symposium, light-activated pesticides were recognized as a new technology with considerable promise in the area of pest control (1). One phototoxin, erythrosin B, has already been registered for house fly control and a promising photodynamic herbicide based on porphyrin metabolism in plants is under development. Among the most active biocides whose activity is enhanced by light are the naturally occurring thiophenes and biosynthetically related polyacetylenes which are characteristic secondary metabolites of the plant family Asteraceae. This particular group of compounds was of sufficient interest to be the focus of a recent specialist conference held in Denmark (2). Many

0097–6156/89/0387–0164$06.00/0

representatives of this group of compounds are phototoxic to both herbivorous insects and/or mosquito larvae. These aspects have been reviewed previously (1,2) and hence the present report will deal with the most active compound, alpha terthienyl (alpha-T) and its derivatives and analogues which have been the subject of recent research.

We have investigated alpha-T extensively as both a potentially useful phototoxic insecticide and as a model phototoxin for plant-insect interaction studies. It is a relatively stable natural product especially when compared to biosynthetically related acetylenes. Alpha-T and related thiophenes are found in many genera including Flaveria, Tagetes, Eclipta, Dyssodia, Nicolletia, Chrysactinia, Adenophyllum at concentrations ranging from 20-440 ug/g dry weight (3,4). When combined with near-UV irradiation (300-400 nm) or natural sunlight, it is exceptionally toxic to mosquito larvae (LD_{50}=19 ppb for Aedes aegypti), blackfly larvae (LD_{50}= 28 ppb Simulium verecundum) and some herbivorous insects (LD_{50}=10 ug/g for contact application to late instar Manduca sexta or Pieris rapae)(5-7). In addition, tests showed that ten other species in three genera of mosquito larvae, some species of sawflies (Tenthridinidae), cutworms (Noctuidae) and loopers (Geometridae) were also very sensitive. The presence of thiophenes in prominent marginal leaf glands of Porophyllum, at the point of first attack by a Lepidopteran herbivore, their high level of toxicity to some hebivores and our recent demonstration of their elicitation in Tagetes after fungal infection is consistent with the plant defence hypothesis (8). However, the existence of insects relatively insensitive to alpha-T, including the European corn borer, tobacco budworm and Anaitis plagiata, a biocontrol agent for the phototoxic St. John's wort indicates that it is not impossible for some insects to circumvent this type of insecticidal agent.

Development of an efficient large scale synthesis of alpha-T by a Grignard-Wurtz reaction (6) allowed us to conduct much more detailed evaluations of alpha-T as an insecticide. Field trials in a deciduous forest site in Eastern Canada were undertaken on mosquito larvae (Aedes spp.) in natural and simulated pools and effects on non-target organisms monitored. Using alpha-T formulated in ethanol, the ED_{50} for control of mosquito larvae was 50 g/ha, an application rate which had no effect on trout, snails, mayfly and caddisfly larvae, but was toxic to Daphnia and chironomids (9).

A surface spreading formulation and an emulsifiable concentrate have been tested in two years of field trials at both a boreal and a deciduous forest site in Eastern Canada. The emulsifiable concentrate formulation gave reliable control of larvae at applications of 100 g active ingredient (a.i.)/ha and the spreading formulation was effective at 50 g a.i./ha, but had the disadvantage of being slightly toxic to non-target organisms due to the xylene-containing formulation.

While good biological and chemical control alternatives exist for insect pest control, there is a need for the development of new technologies because of the problem of development of resistance in target insect populations (11). The continued pressure for maximal crop yields and for the control of insect vectors of human and livestock disease necessitates a continuing search for new control strategies. Recent work from this group has addressed the questions of whether alpha-T is effective against mosquito vectors of human tropical disease and whether it will be an effective control agent for mosquitos that have developed resistance to conventional pesticides. We still know little about what makes a good phototoxin and so a synthetic program has been initiated to determine if other more effective phototoxic molecules can be made. These materials have also provided a collection of molecular structures to investigate the question of what photochemical, physical or structural properties contribute to the design of an effective phototoxin. Tracer materials have been made and toxicokinetic and O_2 toxicity studies undertaken to determine which potential target species are sensitive or tolerant to the phototoxins and by what mechanism.

Larvicidal activity of alpha-T to *Anopheles gambiae* under tropical conditions

To determine the potential of alpha-T as a larvicidal agent against a disease vector, trials were conducted in Tanzania against the malaria mosquito, *Anopheles gambiae*. All trials were conducted in Muheza town, Tanga region (altitude 200ft) during the period of March 10-April 1, 1987. Tests were conducted in partial shade under deciduous trees during weather conditions that were relatively clear. Water temperatures between 32-35 ^{o}C were recorded. Applications of the emulsifiable concentrate at six application rates were made at 10:00 a.m. to pools containing 20 third or fourth instar larvae. Other procedures have been described in detail previously for Canadian field tests (10).

Mean mortality data for seven replicate trials (Figure 1) show that an almost complete mortality of larvae was observed at 100, 50 or 25 g a.i./ha application. At 12.5 and 6.2 g a.i./ha mortality increased rapidly for the first few hours then increased more slowly thereafter. A probit analysis of the 28 h postapplication data gave a linear transformation and an ED_{50} of 7.45 g a.i./ha or an ED_{90} of 18.9 g a.i./ha. Perhaps because of higher light intensities in the tropics, these efficacy values are substantially better

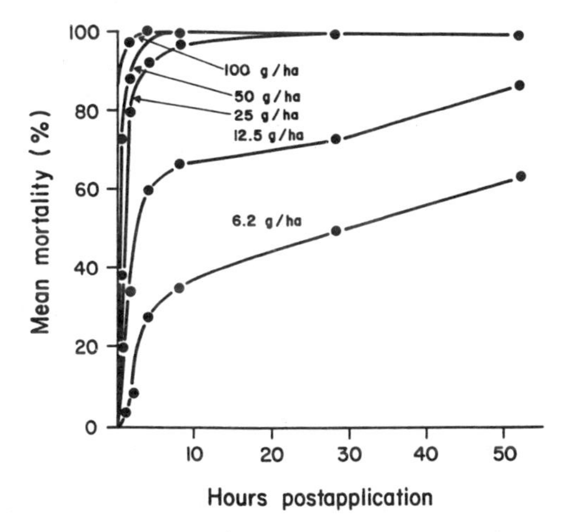

Figure 1. Mortality of fourth instar Anopheles gambae larvae in pools treated with different applications of alpha-terthienyl plotted aginst time after application in hours. Each point is the mean of seven trials with twenty insects per concentration.

than the previously reported Canadian trials where 24 h ED_{50} values ranged from 22.7-93.5 g a.i./ha (10). They compare favourably with results obtained with synthetic pyrethroids (12) and demonstrate the potential of thiophenic phototoxins for mosquito control.

Cross resistance, toxicokinetics and tolerance to phototoxins

To test our hypothesis that the mode of action of phototoxic insecticide is sufficiently different from conventional insecticides to avoid the problem of cross resistance, Hasspieler and Downe examined the toxicity of alpha-T to two strains of Culex tarsalis that vary in their susceptibility to malathion by two orders of magnitude (13). The toxicity to the malathion resistant strain (LC_{50} = 16 ppb) was similar to that found in the malathion susceptible strain (LC_{50} = 12 ppb) (13). The results clearly indicate that a thiophenic phototoxin can be an effective insecticide in populations of insects resistant to another type of material. Similarly, Respicio and Heitz (14) found that there was no cross resistance for the phototoxin erythrosin-B in house flies resistant to several pesticides.

We have previously mentioned that several herbivorous insect species are very tolerant to alpha-T and those that have host ranges that include phototoxic plants are of particular interest. These large differences in sensitivity to alpha-T can be accounted for by different rates of excretion of the phototoxin (table I)(7). Polar metabolites are produced by the insect in which the thiophene conjugation has at least been partially destroyed rendering the molecule non-phototoxic. Studies with the inhibitor piperonyl butoxide indicate that cytochrome P_{450}-based polysubstrate monooxygenases (PSMO) are involved in the metabolism. Examination of the rate of metabolism in the three species per unit of cytochrome P_{450} suggests that there is some evidence of specificity of the system in the most tolerant insect, O. nubilalis, for thiophene metabolism (15). Because this insect has a host range that includes thiophene-producing plants, the specificity may have been selected for by the presence of these toxins in the insect diet.

These studies on the metabolism and excretion of alpha-T by insect herbivores and, more recently, by sensitive mosquitoes were facilitated by the development of radiolabelled phototoxins. ^{3}H-alpha-T and ^{3}H-Me-alpha-T were produced by Werstiuk using a high temperature dilute acid exchange process which labels the aromatic protons on the thiophene rings.

Table I. Sensitivity and PSMO mediated metabolism of alpha-T in three insect herbivores

Species	LD_{50} (ug/g)	alpha-T metabolized (pmoles/min/nmole cytochrome P-450)
Manduca sexta	10	304
Heliothis virescens	474	1538
Ostrinia nubilalis	698	3984

Larvae of Culex tarsalis and Aedes aegypti were exposed to ^{3}H-alpha-T and after absorbtion of the labelled phototoxin, elimination of the labelled compounds was determined by liquid scintillation spectrophotometry. Results showed that rates of elimination of alpha-T differed significantly between the two species and were inversely correlated to the toxic effects of the compound. Over 70% of the label is eliminated in both species within 24 h of exposure and the kinetics of the elimination appear to be polyexponential (13).

Analogue and derivative studies

Using the Grignard-Wurtz reaction, or the Bestmann and Schaper procedure if substitutions on the central thiophene ring were desired, we have now prepared over 30 analogues and derivatives of alpha-T (16). These have been investigated for structure-activity relationships and underlying mechanistic considerations. Scaiano (NRC) has examined the quantum yield of singlet-O_2 generation of a number of these compounds using a laser flash photolysis method (17). The quantum yield of alpha-T itself was an area of some controversy with a very low quantum yield reported by Reyftmann and Kagan (18) and a preliminary report of a quantum yield higher than 1.0 given at the 1986 ACS symposium (1). After repeated measurements, a mean value of 0.86 was determined, indicating that the alpha-T is an extremely efficient singlet-O_2 generator but probably not involved in multiple energy transfer reactions following a single excitation. The quantum yield for 9 other analogues and deriviatives was between 0.53-0.93, indicating that they also are good photo-oxidants. The variation in singlet oxygen quantum yields is small and cannot explain the large variation in the toxicity of the analogues. However, there is a small positive correlation between quantum yield and phototoxicity to mosquito larvae. As has been found for other classes of insecticides, there is also a positive correlation between partition coefficient of the compounds and phototoxicity. Our detailed analysis of these trends will be reported elsewhere, but results indicate that other factors must also be important.

We subsequently investigated the substituent effect and structure-function relationships with these compounds. There was no clear relationship between electron withdrawing or donating substituents (or their Hammett constants) and the phototoxicity of the compounds. Results for the electron releasing compounds are shown in table II but both groups contained compounds that were more or less toxic than the parent. With a single carbon substituent phototoxicity decreased in the order: methyl, aldehyde, alcohol, carboxylic acid (table III). In a 2 carbon atom substituent series the following decreasing order of toxicity was found: acetylene, alcohol, carboxylic acid, methoxy (table III).

Table II. Effect of electron releasing groups on toxicity of alpha-terthienyl derivatives

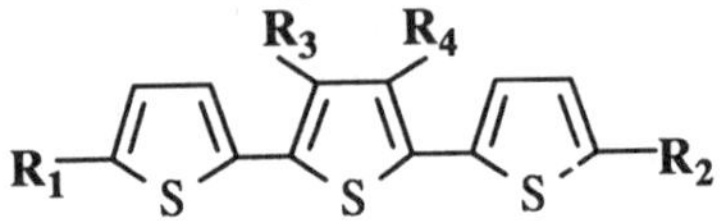

R_1	R_2	R_3	R_4	Relative Toxicity
H	H	H	H	100
Me	H	H	H	200
Me	Me	H	H	8
H	H	Me	Me	18
H	H	Et	Et	3
t-Bu	t-Bu	H	H	>1

Table III. Effect of substituents on toxicity of alpha-terthienyl derivatives

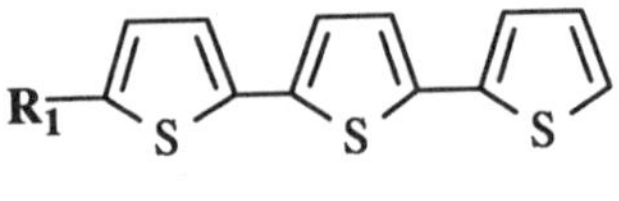

R1	LD_{50} (ppb)	Relative Toxicity
H	30	100
CH_3	15	200
CH_2OH	123	24
CHO	40	75
COOH	1040	3
CH2CH2OH	113	27
$CH2CH_2OCH_3$	1930	2
CH_2COOH	496	6
C≡CH	59	50

Conclusion

The current information clearly indicates that thiophenic phototoxins have advantages in the following area: exceptional activity to the malaria mosquito, Anopheles gambiae, and lack of cross resistance with other pesticides. We have a better understanding of how and why some herbivorous insects may be tolerant through efficient PSMO detoxification systems leading to excretion. It is clear that many alpha-T derivatives and analogues have high quantum yields for production of toxic singlet O_2 which contributes to their toxicity. However toxicity is not fully predictable based on singlet O_2 yields, partition coefficients or Hammett constants and complex biological interactions between phototoxin and target undoubtedly play a role in determining the level of phototoxicity.

Acknowledgements

This research was financed by grants from NSERC (strategic and operating program). We thank Dr. J. Lines for assistance with the Tanzanian trials.

Literature cited

1. Heitz, J.R.; Downum, K.R.(eds.) Light-Activated Pesticides; A.C.S. Symposium series No.339.; American Chemical Society, Washington, D.C. 1987.
2. Lam, J.; Breteler, H.; Hansen, L.; Arnason, J.T. (eds.) Naturally-Occuring Acetylenes; Elsevier, 1988.
3. Downum, K.R.; Kiel, D.J.; Rodriguez, E. Biochem. Syst. Ecol. 1986, 13, 109-113
4. Arnason, J.T.; Morand, P.; Reyes, I.; Lambert, J.D.H.; Towers, G.H.N. Phytochem. 1983, 22, 594-595
5. Arnason, J.T.; Swain, T.; Wat, C.K.; Graham, E.; Partington, S.; Towers, G.H.N.; Lam, J. Biochem. Syst. Ecol. 1981, 9, 63-68
6. Philogene, B.J.R.; Arnason, J.; Berg, C.W.; Duval, F.; Champagne, D.; Taylor, R.G.; Leitch, L.; Morand, P. J. Econ. Ent. 1985,78, 121-126
7. Iyengar, S.; Arnason, J.T.; Philogene, B.J.R.; Morand, P.; Werstiuk, N.H.; Timmins, G. Pesticide Biochem. Physiol. 1987, 29, 1-9
8. Kourany, E.; Arnason, J.T.; Schneider, E. Physiol. Molec. Pl. Pathol. 1988, in press
9. Philogene, B.J.R.; Arnason, J.T.; Berg, C.W.; Duval, F.; Morand, P. J. Chem. Ecol. 1986, 12, 893-898
10. Arnason, J.T.; Philogene, B.J.R.;Duval, F.; Iyengar, S.;Morand, P. in Naturally-Occuring Acetylenes, Lam, J.; Breteler, H.; Hansen, L.; Arnason, J.T. (eds) Elsevier, 1988

11. Georghiou, G.D.; Mellon, R.B. in Pest Resistance to Pesticides Georghiou, G.D.; Saito, T. (eds.) 1983, 1-46, Plenum
12. Helson, B.; Surgenor, G. J. Amer. Mosq. Contr. Assoc. 1986, 2, 269-260
13. Hasspieler, B.; Arnason, J.T.; Downe, A.E.R. J. Amer. Mosquito Control Association. 1988 in press.
14. Respicio, N.C.; Heitz, J.R.; J.Econ. Ent. 1986, 19, 315-317
15. Iyengar, S; Toxicokinetics and Metabolism of Alpha -terthienyl in Manduca sexta, Heliothis virescens and Ostrinia nubilalis Ph.D. thesis, University of Ottawa, 1988
16. MacEachern, A.; Soucy, C.; Leitch, L.; Arnason, J.T.; Morand, P. Tetrahedron 1988, 24, 2403-2408
17. Scaiano, J.C.; MacEachern, A.; Arnason, J.T.; Morand, P.; Weir, D. Photochem. Photobiol. 1987, 46, 193-199.
18. Reyftmann, J.P; Kagan, J.; Santus, R.; Moliere, P. Photochem. Photobiol. 1985, 41, 1-7

RECEIVED November 2, 1988

Chapter 13

Insecticidal Unsaturated Isobutylamides

From Natural Products to Agrochemical Leads

Masakazu Miyakado, Isamu Nakayama[1], and Nobuo Ohno

Pesticides Research Laboratory, Takarazuka Research Center, Sumitomo Chemical Company, Ltd., 4–2–1, Takatsukasa, Takarazuka, Hyogo 665, Japan

For several years, our research group at Sumitomo has been conducting an extensive search for new biologically active natural products. In this chapter, a series of studies on unsaturated N-isobutylamide insecticides is described. An extract of black pepper (Piper nigrum) exhibited strong insecticidal activities against several insects. From the extract, an amide, N-isobutyl-11-(3,4-methylenedioxyphenyl)-(2E,4E,10E)-2,4,10-undecatrienamide (pipercide) and two structurally related amides were isolated as insecticidal principles. In the course of synthetic modifications, N-isobutyl-12-(3-trifluoromethylphenoxy)-(2E,4E)-2,4-dodecadienamide was found to have potent activity. This amide, as well as the amides from pepper plant, exhibited notable paralyzing effects and lethal activity against susceptible and pyrethroid-resistant insects. Electrophysiological studies using the central nerve cord of the American cockroach demonstrated that these amides are neurotoxic. Related synthetic studies of other groups are also described.

From plants of Compositae and Rutaceae, a number of N-isobutylamides of unsaturated and C_{10}-C_{18} acids have been obtained as insecticidal substances. Pellitorine and its analogues provide typical examples (for details, see (1) and (2)).

The occurrence of pellitorine [I] was first described in 1895 by Dunstan et al. (3) as pungent principle in the roots of the medicinal plant, Anacyclus pyrethrum DC. (Compositae). After a long and complicated study, the structure of pellitorine was deduced to possess a conjugated dienamide chromophore (-C=C-C=C-CONH-). It was only in 1952 that the correct structure was confirmed by synthesis to be N-isobutyl-(2E,4E)-2,4-undecadienamide (4). Jacobson reported

[1]Current address: Plant Protection Division—Domestic, Sumitomo Chemical Company, Ltd., Kitahama, Higashi-ku, Osaka 541, Japan

0097–6156/89/0387–0173$06.00/0

that the activity of pellitorine [I] on houseflies (*Musca domestica*) was about one-half that of the pyrethrins (5). Following the structral elucidation of pellitorine, a number of *N*-isobutylamides of unsaturated and aliphatic acids were isolated from many plant species (Figure 1) (2). Interestingly, many of these unsaturated amides of plant origin were reported to exhibit various kinds of biological activities such as: insecticidal (6), molluscicidal (7), pungent (tongue-numbing) (8), coronary vasodilating (9), anti-oxidative (10), pyrethrum synergistic (11), interceptive (postcoital antifertility) (12), anti-tubercular activity (13), and so on. Thus, pellitorine and its related amides could be looked upon as good structural leads for the development of a new class of insecticides or pharmaceuticals.

Since the structural elucidation of pellitorine, a number of analogues were synthesized to develop more potent compounds (14). Some of the synthetic amides were reported to exhibit rapid paralyzing action (knockdown) and lethal toxicity against houseflies, yellow mealworms (*Tenebrio molitor*) and mustard beetles (*Phaedon cochleariae*) (Figure 2) (2). However, these synthetic analogues had irritant properties to human skin as well as toxicity to mammals. In addition to these undesirable side-effects, these unsaturated amides were quite unstable. Every application trial of pellitorine [I] and its synthetic analogues as insecticides gave poor results because of this low stability.

Isolation and structural elucidation of insecticidal constituents from black pepper.

In a search of natural sources for new agrochemical leads with good environmental properties, we chose food-spices as a promising starting point. We expected that spices would bring leads with low mammalian toxicity. Among the 30 spices investigated, only black pepper extract (*Piper nigrum*) exhibited high insecticidal activities against common mosquito larvae (*Culex pipiens pallens*) and adzuki bean weevils (*Callosobruchus chinensis*).

The fruits of *Piperaceae* plants have been known to contain many physiologically active principles, and a number of studies on the chemical constituents of the fruits have been conducted. Among these components, unsaturated amides constitute a major group of secondary metabolites.

Hervil *et al*. reported in 1943 that piperine [II], a major hot principle of black pepper, was more toxic than pyrethrum against houseflies (15). As described earier in this paper, Jacobson reported that pellitorine [I], a plant constituent of *Compositae*, *Rutaceae*, and also obtained from *Piperaceae*, had a notable knockdown effect on houseflies (5). Later, Su (16) and Scott *et al*. (17) reported that crude and purified extracts of black pepper caused high mortality against grain pests such as rice weevils (*Sitophilus oryzae*) and boll weevils (*Anthonomus grandis*) on topical application.

In addition to this, there are many studies on the chemical constituents of *Piperaceae* plants. Over fifty unsaturated amides have been isolated from plant sources, so far (18). However, most of these amides were not studied with regard to insecticidal activity, so the insecticidal nature of these amides from Piperaceae remained obscure.

In 1979, Miyakado et al. reported the isolation of pipercide [IIIa] (19), dihydropipercide [IIIb] and guineensine [IIIc] (20) as genuine insecticidal principles from the fruits of black pepper (Figure 3). Adzuki bean weevils were used as test insects during the isolation process. The structures of pipercide [IIIa] and its analogues [IIIb] and [IIIc] were confirmed by synthesis (20, 21). Until the isolation of [IIIa], [IIIb] and [IIIc], piperine [II] had been said to be a major insecticidal component of black pepper (22). In our observations, piperine itself showed no lethal activity against houseflies, although it exhibited notable synergisms with pellitorine or pyrethrins (23).

Insecticidal activity of Piperaceae amides [IIIa], [IIIb] and [IIIc].

The insecticidal activity of [IIIa], [IIIb] and [IIIc] against adzuki bean weevils is summarized in Table I. Among these amides, dihydropipercide was most toxic; guineensine and pipercide came next, respectively. Interestingly, a mixture of pipercide and the two amides exhibited notable joint action. In particular, a mixture of the amides [IIIa], [IIIb] and [IIIc] in 1 : 1 : 1 ratio demonstrated the highest joint action (two to five fold increase in activity, which was comparable to that of pyrethrins). The knockdown times of the three amides against adzuki bean weevil are also given in Table I (24). When pipercide [IIIa] was applied topically, the knockdown activity (KT50) was observed at 11.8 min. However, dihydropipercide and guineensine exhibited longer knockdown times than pipercide. In this case also, the highest knockdown activity was observed in a mixture of the amides [IIIa], [IIIb] and [IIIc] at a ratio of 1 : 1 : 1. As a reference, the knockdown activity of pyrethrins was examined. The KT50 of pyrethrins was only 1.0 min at the dosage of 0.05 μg/insect. It is noteworthy that the order of knockdown activity is the reverse of the order of lethal toxicity for the three amides. In 1981, Su et al.

Table I. Insecticidal activity and knockdown time of Piperaceae amides to adult C. chinensis (male) on topical application

compds.	LD50 (μg/insect)[a]	KT50 (min)[b]
pipercide [IIIa]	0.56	11.8
dihydropipercide [IIIb]	0.23	30.0
guineensine [IIIc]	0.36	20.5
[IIIa],[IIIb],[IIIc], 1:1:1	0.11	6.0
pellitorine [I]	6.46	>60.0
piperine [II]	>10.0	>60.0
pyrethrins	0.10	1.0[c]

a) Mortalities were evaluated after 24 hr. b) 0.1 μg/insect was applied. c) 0.05 μg/insect was applied.

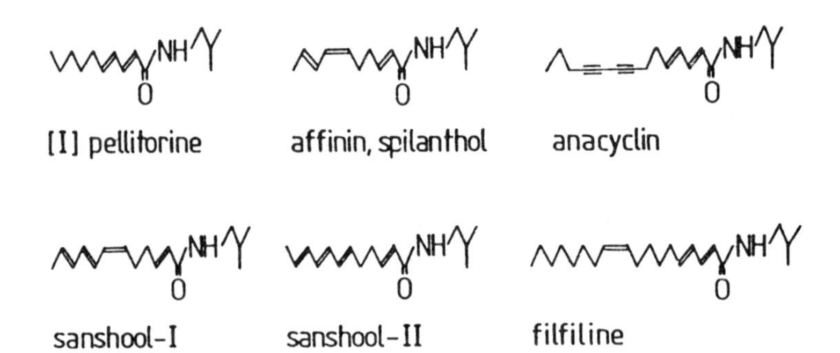

Figure 1. Representative aliphatic unsaturated isobutylamides of plant origin.

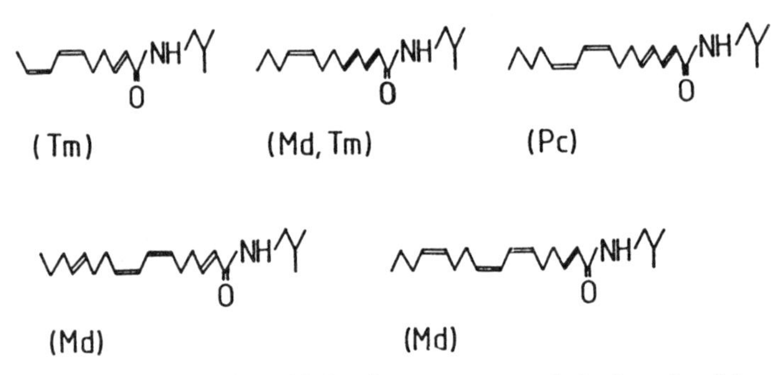

Figure 2. Synthetic aliphatic unsaturated isobutylamides with insecticidal activity.

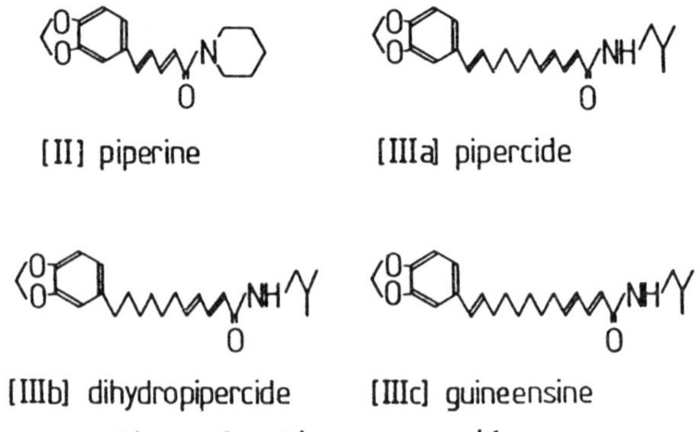

Figure 3. <u>Piperaceae</u> amides.

also studied the constituents of black pepper and reported the isolation of the following amides as insecticidal ingradients: pellitorine [I], pipercide [IIIa] and guineensine [IIIc] (25). The reported insecticidal toxicities (LD50, 24 hr) of [I], [IIIa] and [IIIc] against cowpea weevil (Callosobruchus maculatus, male) were 2.18, 0.25 and 0.84 μg/insect, respectively. These biological observations are concordant with our data in Table I.

Mode of action of pipercide [IIIa].

As described above, pipercide [IIIa] and its analogues [IIIb] and [IIIc] exhibited notable lethal and knockdown activities against adzuki bean weevils. The toxicity of these amides was almost comparable to that of the pyrethrins. Since the discovery of [IIIa], [IIIb] and [IIIc], several biological studies have been conducted to examine insecticidal characteristics and mode of action.

Knockdown activity: First, the knockdown activity of pipercide [IIIa] against American cockroaches (Periplaneta americana) was investigated (26). The knockdown activity of pipercide, pyrethrin I and DDT is shown in Table II. Neither pipercide nor DDT showed knockdown activity within 120 min when 10 μg of these compounds were topically applied to the thorax. On the contrary, pyrethrin I exhibited a KT50 of 12.6 min (10 μg treatment). On the other hand, pipercide showed a typical knockdown effect by the injection method. Burt and Goodchild have pointed out that the effects of pyrethroids depend on the rate of their penetration and arrival at the site of action (27). They also mentioned that penetration of pyrethroids affected knockdown activity but had no influence on lethal toxicity.

Table II. Knockdown activity against P. americana via abdominal injection or topical application to thorax at 10 μg/male

compds.	KT50 (min)[a]	
	injection	topical
pipercide [IIIa]	10.0	>120
pyrethrin I	4.4	12.6
DDT	>120	>120

a) Only initial KT50 times are given in this table. Several recovered insects were observed after initial knockdown.

As expected, pyrethrin I exhibited more rapid knockdown action by the injection method than by topical application. Similarly, pipercide was more effective when treatment was by the injection method than by topical application. Therefore, it could be assumed that pipercide had a slower penetration rate than that of pyrethrins. From these observations, it was estimated that pipercide did not show knockdown activity owing to its slow penetration. Therefore, pipercide exhibited fast knockdown actionwhen a proper amount of it existed in the insect body.

Synergism:
Next, the effect of synergists on pipercide activity [IIIa] was studied. German cockroaches (Blattera germanica, male) were used as test organisms (Umeda, K.; Miyakado, M., Sumitomo Chem. Co. Ltd., unpublished data). PBO (100 μg/insect) or NIA-16388 (28) (20 μg/insect) was applied topically 2 hr before the injection of pipercide. The lethal toxicity of pipercide without synergist was 5.00 μg/insect (LD50, 24 hr). However, the toxicity of [IIIa] was synergistically enhanced to 1.55 and 0.51 μg/insect, respectively, when the insects were pre-treated with PBO or NIA-16388. This result suggested that pipercide had relatively nonresistant moieties with respect to the action of oxidative or hydrolytic enzymes.

Effect on nervous system:
The effect of pipercide [IIIa] on the central nerve cord system (CNS) of P. americana was investigated (26). Pipercide induced a spike increase at 2.5 min and repetitive discharge at 2.5-3.0 min, while conduction blockage was not observed within 30 min. On the other hand, pyrethrin I caused a spike increase at 0.3 min and repetitive discharges at 0.3-1.0 min as a rapid response. Conduction blockage was also observed on the CNS at 15 min. DDT exhibited no effect on the CNS within 30 min. Electrophysiological studies showed that pipercide acted on the CNS at 10^{-5}M: the same as pyrethrin I but to a lesser degree.

Electrophysiological studies using a pyrethroid-resistant American cockroach CNS preparation revealed the same response as that of susceptible organisms. Pipercide exhibited the same order of toxicity when 5.0 μg/insect was injected to susceptible or pyrethroid-resistant American cockroaches. Although the poisoning symptoms of pipercide were quite similar to those of the type I pyrethroids, it was estimated that pipercide might act at a different site within the nervous system.

Synthesis of pipercide [IIIa].

The first synthesis of pipercide was reported by our group in 1979 (21). The synthetic scheme is given in Figure 4. One of the hydroxy groups of 1,6-hexanediol [IVa] was selectively protected as a pyranylether, then the other was oxidized with pyridinium chlorochromate to give 6-(tetrahydro-2'-pyranyloxy)-1-hexanal [IVb]. The aldehyde [IVb] was subjected to Wittig-Wadsworth-Emmons reaction, using diethyl 3-methoxycarbonyl-2-propenylphosphonate [IVc] in the presence of NaOMe, yielding the (2E,4E)-conjugated diene ester [IVd]. The unsaturated ester [IVd] afforded methyl 10-oxo-(2E,4E)-2,4-decadienoate [IVe] on successive treatment with p-toluenesulfonic acid and subsequent oxidation with pyridinium chlorochromate. The 10-oxo-ester afforded the condensation product ((E)-rich) [IVf] by Wittig reaction of 3,4-methylenedioxy-benzyltriphenylphosphonium bromide in benzene in the presence of an equimolar amount of n-BuLi. This conjugated ester [IVf] gave the corresponding acid on hydrolysis with KOH-MeOH. The acid was recrystallized twice from benzene to give pure 11-(3,4-methylenedioxyphenyl)-(2*E*,4*E*,10*E*)-2,4,10-undecatrienoic acid. The acid was converted to the corresponding acid chloride and finally condensed with isobutylamine to give *N*-isobutyl-11-(3,4-methylenedioxyphenyl)-(2*E*,4*E*,10*E*)-2,4,10-undecatrienamide [IIIa]. The synthetic material was identical in all respects to natural pipercide.

The syntheses of dihydropipercide [IIIb] and guineensine [IIIc] were also completed according to almost same procedure as that involved in the pipercide synthesis (20).

Okwute *et al.* (29) and Vig *et al.* (30) have achieved synthesis of guineensine independently and almost at the same time.

Since the first total synthesis of pipercide, several facile and convenient synthetic routes to pipercide have been reported by several groups. Crombie and Denman's synthesis (31) of pipercide was characterized by stereospecific introduction of the (10*E*)-olefinic bond. They employed a dehydration reaction of 1-(3,4-methylenedioxyphenyl)-7-tetrahydro-2'-pyranyloxy)-1-heptanol with methyltriphenylphosphonium iodide in HMPA under heating conditions. They reported the (*E*)-olefin thus obtained was virtually stereoisomerically pure.

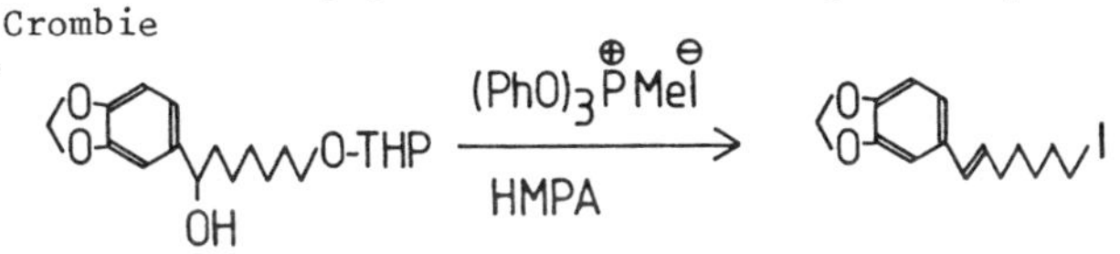

Bloch and Hassan-Gonzales's synthesis (32) of pipercide was characterized by a highly stereospecific preparation of the (*E*,*E*)-conjugated dienoates. They applied a cheletropic reaction with thermal SO_2 extrusion from *cis*-2,5-dialkyl-2,5-dihydrothiofene-1,1-dioxides generated by a retro Diels-Alder reaction. The (*E*,*E*)-dienoate thus obtained has been reported to be of 95% stereoisomeric purity.

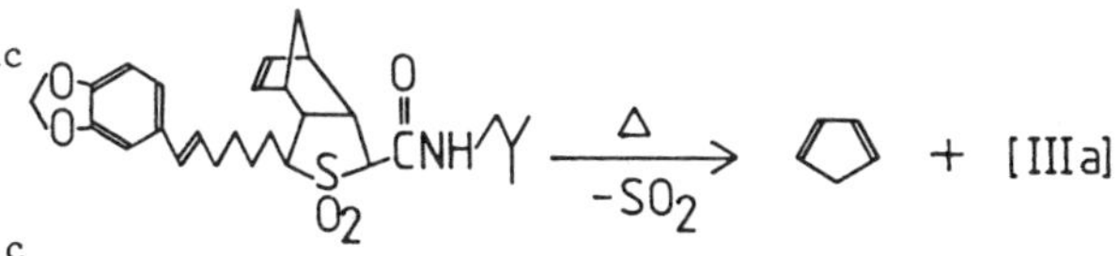

In addition to those syntheses described above, there are many reports on pellitorine analogue ((2*E*,4*E*)-2,4-dienamide) synthesis. As an example, Mandai *et al.* reported (2*E*,4*E*)-conjugated dienoate synthesis from alkyl-SO_2Ph via a double elimination reaction of β-acetoxy sulfone with good selectivity (33). This might be a generally applicable synthetic method for (*E*,*E*)-conjugated dienamides.

Structural modifications.

As described, three *Piperaceae* amides [IIIa], [IIIb] and [IIIc] showed distinct insecticidal and physiological characteristics as summarized hereafter (34):

(a) rapid knockdown action
(b) high toxicity against pyrethroid resistant pests
(c) enhanced toxicity by mixing [IIIa], [IIIb] and [IIIc] (joint action)

(d) increased molecular stability in comparison to that of pellitorine

The chemical structures of these amides [IIIa], [IIIb] and [IIIc] provide valuable information in designing more active synthetic analogues. Among these amides, dihydropipercide [IIIb] was most toxic to adzuki bean weevils, so [IIIb] was chosen as the key structural lead for synthetic modifications. The structural similarities and differences between pellitorine [I] and dihydropipercide [IIIb] were quite suggestive. Although both molecules have a common structural unit (Part A and Part B: the N-isobutyl-(2E,4E)-2,4-dienamide moiety, see Figure 5), the insecticidal activity of the former was far inferior to that of the latter (Table I).

As described in the previous article (24), pellitorine [I] is quite unstable compound. However, with the introduction of a phenyl ring in the molecule, [IIIb], as well as [IIIa] and [IIIc], exhibited improved stability. These observations encouraged us to suspect that by changing the structure, one might be able to improve the insecticidal activity of this class of compounds. In the corresponding structural modification studies, adzuki bean weevils (adult, male) were used as test insects. As shown in Figure 5, the key molecule [IIIb] was divided into three parts. Details of these syntheses were given in our previous report (35) and (36).

Part A:

Our first study was conducted on the amine moiety. The results are summarized in Table III. Among them, the natural amide [IIIb] was most toxic (0.23 μg/insect). By changing the amide moiety from isobutylamine to other branched or cyclic aliphatic amines [Vb,c,d], their toxicities decreased by one-third or one-fourth compared with that of [IIIb]. Only 1,2-dimethylpropylamide [Va] exhibited the same order of activity as that of [IIIb]. Investigation of other amines, such as aniline or benzylamine [Vf,g] or the oxygen containing amine [Ve], gave only poor results.

These results were in good agreement with our previous study in which insecticidal toxicities of pellitorine [I] and its amine analogues were investigated. In that communication, we reported that pellitorine (N-isobutylamide) was the most toxic amide and N-cyclohexylamide was next. Recently, Elliott *et al.* also investigated the adaptability of various amines for different types of unsaturated acids in connection with insecticidal activities (37). They selected several branched aliphatic amines, with four to six carbon atoms, as the appropriate components. These results are comparable with our previous observations. In short, it was concluded that there is a steric limitation on the size of the amine in relation to the corresponding insecticidal activity.

Part B:

As a next step, our study was focussed on part B; i.e., while the N-isobutylamide and 3,4-methylenedioxyphenyl moieties were kept

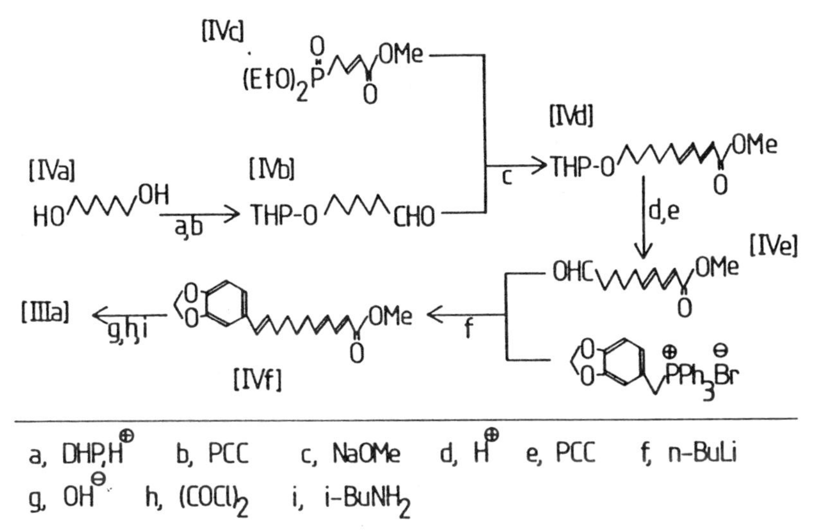

Figure 4. Synthesis of pipercide [IIIa].

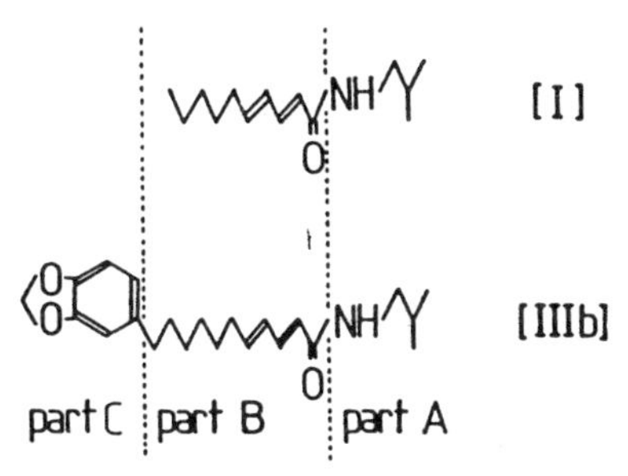

Figure 5. Structure of pellitorine [I] and dihydropipercide [IIIb].

Table III. Insecticidal activities of 11-(3,4-methylenedioxyphenyl)-(2E,4E)-2,4-undecadienamides with various amine moieties to adult *C. chinensis* (male)

	R (amine)	LD50 (μg/insect)[a)]
[IIIb]	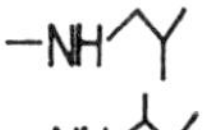 —NH	0.23
[Va]	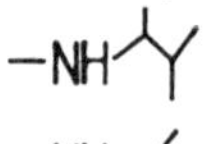 —NH	0.32
[Vb]	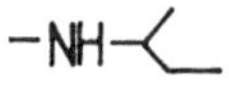 —NH	1.05
[Vc]	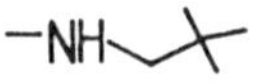 —NH	0.81
[Vd]	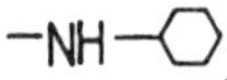 —NH	1.00
[Ve]	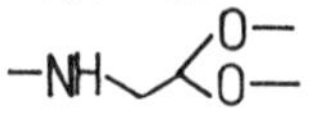 —NH	>10
[Vf]	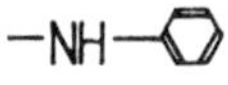 —NH	>10
[Vg]	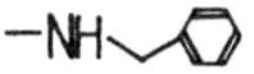 —NH	>10

a) topical application (mortalities were evaluated after 24 hr). (Reproduced with permission from Ref. 35. Copyright 1985, Pesticide Science Society of Japan.)

unchanged, the suitability of conjugated olefins, chain length and the introduction of an ether group in a straight chain were examined (Table IV). The saturated or monoene amide [VIa,b] exhibited no toxicity. By introducing a 3,4-methylenedioxyphenoxy group [VId] into the molecule, instead of the corresponding benzyl group [VIc], almost no change in their activity was observed. However the phenoxy analogues were superior to the benzyl analogues in view of the simplicity of their synthesis. Introduction of a methyl group at C-3 in the (2E,4E)-dienamide moiety [VIe] showed marked enhancement of activity. In contrast, compound [VIf], with a (2Z,4E)-dienamide, exhibited negligible toxicity.

Part C:

The effect of the aromatic ring substituents on insecticidal activity was studied with the phenoxy analogues (Table V). It turned out that the 3,4-methylenedioxyphenyl group, a common substituent of *Piperaceae* amides, was not essential for toxicity. 4-Halogen-substituted analogues [VIId,e] exhibited ten times the toxicity of the corresponding methylenedioxy analogue [VId]. 3-Halogen-substituted analogues, especially, 3-Br [VIIh], as well as the 3-Cl [VIIg] and 3-CF_3 [VIIi] analogues, exhibited the highest toxicity. These amides were fifty to eighty times more toxic than the natural product, dihydropipercide [IIIc]. The 4-CH_3, 4-NO_2 and 4-OCH_3 derivatives ([VIIa], [VIIb] and [VIIc], respectively) revealed almost no toxicity. It is noteworthy that the amides

Table IV. Insecticidal activities of N-isobutyl-(3,4-methylenedioxyphenylamides with various carbon chain length to adult C. chinensis (male)

compds.	LD50 (μg/insect)
 [IIb]	0.23
[VIa]	 >10
[VIb]	 >10
[VIc]	 1.80
[VId]	 2.00
[VIe]	 0.38
[VIf]	 >10

MDP : 3,4-methylenedioxyphenyl, iBu : isobutyl.
(Reproduced with permission from Ref. 35. Copyright 1985, Pesticide Science Society of Japan.)

[VIIh,i] showed the same order of activity as that of fenitrothion, one of the well known organo phosphorous insecticides.

Table V. Insecticidal activities of N-isobutyl-12-(substituted phenoxy)-(2E,4E)-2,4-dodecadienamides to adult C. chinensis (male)

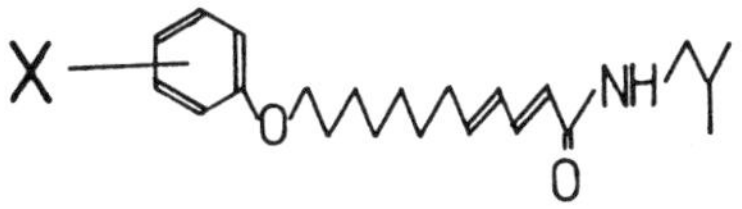

compds.		LD50 (μg/insect)
[VId]	X = 3,4-methylenedioxy	2.00
[VIIa]	4-CH_3	>10
[VIIb]	4-NO_2	8.00
[VIIc]	4-OCH_3	>10
[VIId]	4-Cl	0.15
[VIIe]	4-Br	0.25
[VIIf]	3,4-diCl	0.070
[VIIg]	3-Cl	0.068
[VIIh]	3-Br	0.038
[VIIi]	3-CF_3	0.043
	pyrethrins	0.10
	fenitrothion	0.045

(Reproduced with permission from Ref. 36. Copyright 1985, Pesticide Science Society of Japan.)

According to these structural modifications, N-isobutyl-12-(3-trifluoromethylphenoxy)-(2E,4E)-2,4-dodecadienamide [VIIi] was selected as the eminent structure. Compound [VIIi]

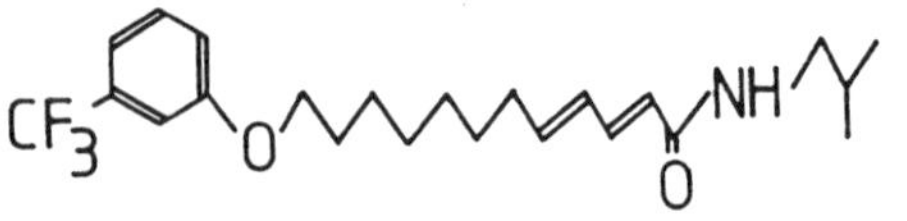

[VIIi]

exhibited strong lethal activity against adzuki bean weevills as well as rice stem borers (Chilo suppressalis) and houseflies. However, this amide did not show any toxicity against tobacco cutworms (Spodoptera litura). Further investigations must be performed on the relationships between structure and insect-spectrum of this amide series.

Recently, several groups have reported structural modification research.

Black et al. selected N-isobutyl-9-(3-trifluoromethylbenzyloxy)-(2E,4E)-2,4-nonadienamide as one of the most potent amides against houseflies (37). The structure of this amide was close to our amide [VIIi] and the reported insecticidal activity of this amide was also similar to [VIIi].

Elliott and his group conducted their work on different types of unsaturated amides (38). They selected two amides: N-(2,2-dimethylpropyl)-6-(3,5-difluorophenyl)-(2E,4E)-2,4-hexadienamide against houseflies and N-isobutyl-6-(3,4-dibromophenyl)-(2E,4E)-2,4-hexadienamide against mustard beetles, as potent amides, respectively. They also studied the insecticidal action of phenylhexadienamides on a pyrethroid-resistant (super-kdr) strain of houseflies (39). As expected, these amides exhibited lethal and knockdown activities against the R-strains as well as the S-strains. Synergists were also effective in increasing the insecticidal activities of these phenylhexadienamides (39).

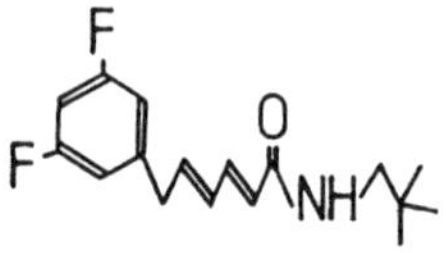

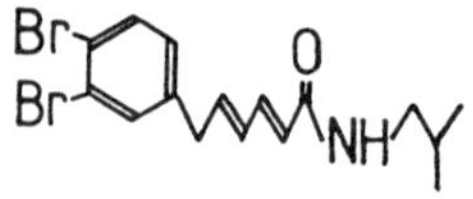

Much progress have been achieved since the 1979 discovery of pipercide [IIIa] as a new lead for insecticidal amides. However, further investigation must be conducted for practical development of this class of compounds as new insecticides.

To state these goals clearly, (i) More lipophilicity must be given to the molecule to increase the penetration rate through the insect cuticle. (ii) Insecticidal potency must be improved as much as possible. (iii) Mode of action must be understood (especially, against pyrethroid-resistant strains). (iv) The conformations of unsaturated amides (in solution) must be clarified.

Continued synthetic work is currently in progress and will be reported in due course.

Concluding Remarks.

In this chapter, we described the progress of search on the

insecticidal unsaturated N-isobutylamides --- from the natural pellitorine era to Piperaceae amides --- including isolation, structure elucidation, mode of action, synthesis and structural modifications. Most of the potent insecticides are modeled on natural products. They include esfenvalerate, based on the pyrethrins of chrysanthemum flowers, and caltap, modeled on a marine worm toxin, nereistoxin. This research on insecticidal unsaturated amides is still in progress, and future discoveries will certainly bring this type of amide to market. We believe that the present research offers an innovative example of how natural product chemistry can provide new leads for tomorrow's agrochemicals. To this end, the present chapter should be read in conjunction with references (1) and (2).

Acknowledgment.

We wish to thank to Dr. H. Yoshioka of the Institute of Physical and Chemical Research, Japan and Prof. N. Nakatani of the Faculty of Science of Living, Osaka City University, Japan for their valuable advice. We are indebted to Dr. David P. Richardson of the Department of Chemistry, Williams College, MA, USA for his critical reading of the manuscript. Finally, the authors wish to thank to Dr. Y. Nishizawa of the Sumitomo Chemical Co. Ltd., for encouragement.

Literature Cited.

1. Jacobson, M. In Naturally Occurring Insecticides; Jacobson, M., Crosby, D. G., Ed.; Marcel Dekker: New York, 1971; p 137.
2. Su, H. C. F. In Comprehensive Insect Physiology Biochemistry and Pharmacology; Kerkut, G. A., Gilbert, L. I., Ed.; Pergamon Press: Oxford, 1985; Vol. 12, p 603.
3. Dunstan, W. R.; Garnett, H. J. Chem. Soc. 1895, 67, 94.
4. Crombie, L. Chem. Ind. (London) 1952, 2997.
   Jacobson, M. J. Amer. Chem. Soc. 1953, 75, 2584.
5. Jacobson, M. J. Amer. Chem. Soc. 1949, 71, 366.
6. LaLonde, R. T.; Wong, C. F.; Hofstead, S. J.; Morris, C. D.; Gardner, L. C. J. Chem. Ecol. 1980, 6, 35.
   Jondiko, I. J. O. Phytochem. 1986, 25, 2289.
7. Kubo, I.; Klocke, J. A.; Matsumoto, T.; Kamikawa, T. In Pestic. Synth. Ration. Approaches; ACS Symposium Series No. 255; American Chemical Society: Washington, DC, 1984; p 163.
8. Yasuda, I.; Takeya, K.; Itokawa, H. Chem. Pharm. Bull. 1981, 29, 1791.
   Greger, H.; Hofer, O. Phytochem. 1984, 23, 1173.
9. Shoji, N.; Umeyama, A.; Saito, N.; Takemoto, T.; Kajiwara, A., Ohizumi, Y. J. Pharm. Sci. 1986, 75, 1188.
10. Nakatani, N.; Inatani, R.; Ohta, H.; Nishioka, A. Environ. Health Perspect. 1986, 67, 135.
11. Atal, C. K.; Dhar, K. L.; Gupta, O. P.; Gupta, S. C.; Saxena, B. P.; Koul, O. Indian J. Exp. Biol. 1977, 15, 1230.

12. Chandhoke, N.; Gupta, S.; Dhar, S. Indian J. Pharm. Sci. 1978, 40, 113.
13. Gupta, O. P.; Nath, A.; Gupta, S. C.; Srivastava, T. N. Bull. Med.-Ethno-Bot. Res. 1980, 1, 99.
14. Weed, A. Soap. Sanit. Chem. 1938, 14, 133.
Crombie, L.; Shah, J. D. J. Chem. Soc. 1955, 4244.
Dominguez, J. A.; Leal Diaz, G.; de los A. Vinales, D. M. Ciencia (Mexico) 1957, 17, 213.
Crombie, L.; Manzoor-i-Khuda, M. J. Chem. Soc. 1957, 2767.
Crombie, L.; Krasinski, A. H. A.; Manzoor-i-Khuda, M. J. Chem. Soc. 1963, 4970.
Meisters, A.; Wailes, P. C. Aust. J. Chem. 1966, 19, 1207.
15. Harvil, E. K.; Hartzell, A.; Arthur, J. M. Contrib. Boyce Thompson Inst. 1943, 13, 87.
16. Su, H. C. F. J. Econ. Entomol. 1977, 70, 18.
17. Scott W. P.; McKibben G. H. J. Econ. Entomol. 1978, 71, 343.
18. Achenbach, H.; Fietz, W.; Woerth, J.; Waibel, R.; Portecop, J. Planta Med. 1986, 12.
19. Miyakado, M.; Nakayama, I.; Yoshioka, H.; Nakatani, N. Agric. Biol. Chem. 1979, 43, 1609.
20. Miyakado, M.; Nakayama, I.; Yoshioka, H. Agric. Biol. Chem. 1980, 44, 1701.
21. Miyakado, M.; Yoshioka, H. Agric. Biol. Chem. 1979, 43, 2413.
22. Windholz, M., Ed.; In The Merck Index (10th Ed.); Merck & Co. Inc.: Rahway, N.J., 1983; p 1077.
23. Miyakado, M.; Nakayama, I.; Yoshioka, H. J. Chem. Soc. Japan 1981, 886.
24. Miyakado, M.; Nakayama, I.; Ohno, N.; Yoshioka, H. In Natural Products for Innovative Pest Management; Whitehead, D. L.; Bowers, W. S., Ed.; Pergamon Press: Oxford, 1983; Chapter 22.
25. Su, H. C. F.; Horvat, R. J. Agric. Food Chem. 1981, 29, 115.
26. Hatakoshi, M.; Miyakado, M.; Ohno, N.; Nakayama, I. Appl. Entomol. Zool. 1984, 19, 288.
27. Burt, P. E.; Goodchild, R. E. Pestic. Sci. 1974, 5, 625.
28. Miyamoto, J.; Suzuki, T. Pestic. Biochem. Physiol. 1973, 3, 30.
29. Okwute, S. K.; Okorie, D. A.; Okogun, J. I. Niger. J. Nat. Sci. 1979, 1, 9.
30. Vig. O. P.; Sharma, S. D.; Sood, O. P.; Bari, S. S. Indian J. Chem. Sect. B. 1980, 19B, 350.
31. Crombie, L.; Denman, R. Tetrahedron Lett. 1984, 4267.
32. Bloch, R.; Hassan-Gonzales, D. Tetrahedron 1986, 42, 4975.
33. Mandai, T.; Moriyama, T.; Tsujimoto, K.; Kawada, M.; Otera, J. Tetrahedron Lett. 1986, 603.
34. Miyakado, M.; Nakayama, I.; Inoue, A.; Ohno, N.; Yoshioka, H. Proc. 5th Intern. Cong. Pesticide Chem. (IUPAC); 1982, Kyoto, Abstract paper, IIIc-28.
35. Miyakado, M.; Nakayama, I.; Inoue, A.; Hatakoshi, M.; Ohno, N. J. Pesticide Sci. 1985, 10, 11.
36. Miyakado, M.; Nakayama, I.; Inoue, A.; Hatakoshi, M.; Ohno, N. J. Pesticide Sci. 1985, 10, 25.
37. Black, M. H.; Blade, R. J.; Moss, M. D. V.; Nicholson, R. A. Proc. 6th Intern. Cong. Pesticide Chem. (IUPAC); 1986, Ottawa, Abstract paper, 1B-21.

38. Elliott, M.; Farnham, A. W.; Janes, N. F.; Johnson, D. M.; Pulman, D. A. Pestic. Sci. 1987, 18, 191 (part 1) to 239 (part 6).
39. Elliott, M.; Farnham, A. W.; Janes, N. F.; Johnson, D. M.; Pulman, D. A.; Sawicki, R. M. Agric. Biol. Chem. 1986, 50, 1347.

RECEIVED November 2, 1988

Chapter 14

Insect Growth Inhibitors from *Petunia* and Other Solanaceous Plants

C. A. Elliger and A. C. Waiss, Jr.

Agricultural Research Service, U.S. Department of Agriculture, 800 Buchanan Street, Albany, CA 94710

From Petunia and Physalis species were obtained a number of steroidal compounds that contribute to plant resistance against insect feeding. Petuniasterones from P. hybrida and the ancestoral species P. parodii and P. axillaris varied in larval growth inhibiting activity toward Heliothis zea depending on structure with the most active materials having a bicyclic orthoester system on the steroid side chain. Physalis peruviana contained numerous withanolides, among which highly glycosylated derivatives appeared to be most active against H. zea. These allelochemicals are discussed in relation to the eventual transfer of the substances into economically significant crops to provide an expanded basis for insect resistance.

For many years chemists have been isolating and identifying substances from plants that are more or less effective in suppressing insects. The investigation of these substances among which are antifeedants, growth inhibitors, antihormones, and other toxicants is justified on the basis of finding on the one hand a new class of "natural pesticides" for commercial application or on the other a relationship between these phytochemicals and "host plant resistance" in appropriate cases. At the present time, however, after all these efforts, only the old well known and established, pyrethrins, rotenoids and nicotine are used as "insecticides" in commercially significant quantities (1). There are, of course, a number of promising candidates whose acceptance would depend upon economic considerations other factors being equal. A difficulty with compounds showing anti-insect activity is that they are not toxic enough, in general, to be commercially satisfactory for external application, and this then leads to the basic question of host plant resistance. Why do these substances confer a selective advantage to the plant? We feel that these agents do provide just this advantage when they are presented by

Published 1989 American Chemical Society

the plant to its herbivores via an efficient delivery system. Generally, dosage of deleterious substances takes place during insect feeding on an otherwise vulnerable plant part or repellency is manifested in close proximity to that part. It is not in principle necessary, therefore, for the plant to cover its entire structure or its vicinity with potent insecticides. It is sufficient for the purpose of defense if exposure to a compound of moderate biological activity be self-inflicted by the attacking insect. For this reason there are very few substances from plants that are potent enough to be termed "insecticides".

A feature of host plant resistance exists that limits its usefulness in crop protection. Such resistance is not actually effective enough against the pests that are of major economic concern. This is not a surprise to the ecologist who realizes that our present observations relate to a co-evolutionary stalemate where certain insects have carved out their own niches by evolving tolerance to substances that may once have been effective against them. Thus, for example, gossypol in cotton serves to ward off attack of many insects but actually attracts the boll weevil (2). Similarly, the cucurbitacins, serving to repel most insects, are kairomones for specialist beetles (3). The attempts by the plant breeder to utilize host plant resistance have been toward increasing the content of active substances already present within a particular crop (4, 5) or to introduce resistance factors found in closely related plant varieties (6). However, the amount of variability available in this manner is limited by the genetic pool of a given plant species and may, in many if not most cases be insufficient to afford protection especially in view of the co-evolutionary process mentioned above. At best, it may be expected that selective breeding would be simply able to restore to a specific crop that degree of resistance which already existed in its wild progenitors. This then is a quantitative effect, but the putative resistance factors will not in these cases differ qualitatively from those to which the pest insects have already developed tolerance. To circumvent the limitations of classical host plant resistance a new means of introducing resistance factors must be found. This means is now available within the techniques of modern cell biology and molecular genetics. We shall eventually be able to provide the essential qualitative transformations of host plant resistance by transferring sources of resistance not merely between species within the same genus but ultimately between genera within a plant family.

We have chosen the tomato plant (*Lycopersicon esculentum*) as an experimental subject upon which to test this concept. The tomato was chosen because of its economic significance and because its complement of existing resistance factors has been well studied (7-9). Additionally, the linkage maps of nuclear chromosomes in tomato are among the most well established (10). As an example of a significant insect pest of this plant, we picked the polyphagous lepidopteran *Heliothis zea* (Boddie) the larvae of which are known synonymously as tomato fruitworm, corn earworm, cotton bollworm and soybean podworm. Experimental advantages of this insect include its ready adaptation to artificial diets and its broad host range which render it a very severe test of potential plant defense systems. Before we could begin to explore advanced genetic techniques it was necesary to screen various candidates among various solanaceous plants that appeared suitable on the basis of other considerations (such as lack of known substances deleterious to humans). We examined a number of species in several genera within the Solanaceae (Table I) in an exploratory screening for activity against *H. zea* and selected a number of active species for further study.

Table I. Species Examined within the Solanaceae for Activity Against *Heliothis zea*

Genus	Species
Lycopersicon	*esculentum*, *hirsutum*, *pervianum*
Solanum	*lycopersicoides*, *tuberosum*, *melongena*
Capsicum	*annuum*
Nicotiana	*tabacum*
Physalis	*alkakengi*, *ixocarpa*, *peruviana*
Petunia	*hybrida*, *axillaris*, *parodii*, *violacea*
Brugmansia	*suaveolens*
Cyphomandra	*betacea*
Salpiglossis	*sinuata*

Table II presents the results of a feeding study using foliage of a few of these species. Four-day old larvae of *H. zea* were placed

Table II. Survival of 4-Day Old *H. zea* Larvae[a] on Several Solanaceous Species

	Percent Survival Days after Application				Weight of Survivors After 6 Days + s.d.	
	1 Day	2 Days	4 Days	6 Days		
L. esculentum (Ace)	100	90	90	90	261 mg	± 120
L. hirsutum, f. glab.	95	90	90	90	103	40
Physalis ixocarpa	95	95	95	95	20	10
P. peruviana	90	90	80	55	6	1
Petunia hybrida (Blue Cloud)	75	50	35	35	90	59
Control Diet[b]	100	100	100	100	700	78

a. 20 larvae initially applied.
b. Artificial diet used in bioassays and insect rearing containing no additive compounds (14).

upon leaves in petri dishes provided with moist filter paper to maintain freshness. As feeding progressed, leaves were changed daily and the larvae were allowed to feed for a total of six days on the leaves (ten days overall). It can be seen that none of the experimental sets had larvae that grew to a size approaching that of control animals. This is a manifestation of the presence of growth inhibiting factors which are known to be present in even the standard susceptible plant tested (Ace tomato) (7-9). This effect is even more pronounced with neonate larvae for which mortality is high and stunting is so severe that individual weights were not practical to obtain. Exposure of plant material to older larvae is a more severe test of plant defenses inasmuch as these older animals are more tolerant in general to allelochemicals (11). The two Physalis species and the Petunia horticultural variety possessed sufficient additional resistance over Ace tomato (Lycopersicon esculentum) to merit consideration as sources of transgenic resistance factors. The other Lycopersicon species, L. hirsutum, F. glabratum is amenable to standard techniques of plant crossing (12). It may be noted that in the latter case, the responsible agent for enhanced resistance, 2-tridecanone, has already been described (13). Physalis and Petunia, of course, differ greatly from tomato in plant characteristics with the former producing a large berry-like fruit within an expanded fused calyx and the latter being known for its flowers. It may be admitted at this time that transgenic combinations of these with Lycopersicon will probably produce a range of plant types which may or may not have desirable properties from an economic standpoint. Our concern is with host plant resistance, however, and we feel that any enhancement of resistance within progeny of these combinations would be of great importance.

We established that the resistance factors in Physalis and Petunia are chemical in nature by serial extraction of leaf material with successively more polar solvents followed by incorporation of material from these extracts into artificial diets. Fractions obtained from extraction of plant material (and eventually solutions of pure compounds) were evaporated onto cellulose powder and incorporated into modified Berger diet (14). Newly hatched larvae of Heliothis zea were applied to each of ten replicate diets and were allowed to feed for ten days. Larval weights were obtained and were compared to the weight of control larvae that were grown on diets containing as additive only cellulose powder. Generally, we have expressed larval growth as precent of control values, and the term ED_{50} (effective dosage) is defined as that concentration of additive required to reduce larval growth to fifty percent of control weights. Thus, we found that active material in Petunia is found in hexane and ethyl acetate fractions (or in chloroform), but not in subsequent extracts of higher polarity. The active substances of the Physalis species are substantially more polar and appear in acetone and methanol extracts. Identification of the respective active components of these extracts is of substantial importance, first in order to follow their appearance in the transgenic progeny and secondly to explore their toxicology.

Petuniasterones. Further fractionation of chloroform extracts from Petunia hybrida in a sequence of chromatographic procedures yielded a large number of steroidal ketones of unusual functionality that vary in their potency toward inhibiting development of H. zea larvae. We have termed these petuniasterones (15) and have categorized them into several groups which are shown here. All of these compounds are derived from a 28 carbon steroid and have a carbonyl group at C-3 with either a hydroxy- or acetoxy- group at C-7. The side chain possesses oxygenated functionality at positions -22, -24, and -25. Among the members of the various series are A-ring dienones and 1-acetoxyenones. Further oxygenation may be found at C-12, C-17 and upon the position α- to the thiolester moiety which is present in some cases as a substituent on a bicyclic orthoester system. The side chain may have orthoester substitution, hydroxy or ester groups flanked by oxirane functionality, three hydroxy groups, or a cyclic ether with a hydroxy substituent. The orthoesters and oxirane containing compounds are much more abundant in plant extracts than the last two substances mentioned. We have observed that facile conversion of one type to another can occur in vitro. Thus, elimination of the 1-acetoxy group in series B, E and H takes place under mild conditions to afford the corresponding dienones. Also, it was noted that the bridged orthoester systems of the A, D and E series may be formed under acidic catalysis from the corresponding petuniasterones B or C whereby the carboxylate ester attached to C-22 is converted to orthoester and the epoxy ring is opened (Equation 1). PS-F, $-G_1$ and $-G_2$ were formed when unesterified PS-C was treated with aqueous acid (Equation 2). We suspect that similar conversions occur naturally within the plant from appropriately substituted 24,25-oxirane containing precursors.

As mentioned above the petuniasterones differ in inhibitory activity toward H. zea. Tables III and IV show the results of feeding studies in which dietary levels of added petuniasterones were presented to larvae at concentrations up to 800 ppm.

Table III. Growth Inhibitory Activity of Petuniasterone Toward Heliothis zea

Compound	MW	PPM	Micromoles/Kg
Petuniasterone A (I)	558	130	233
Petuniasterone A Acetate	600	144	237
Methyl Ester from PS-A	542	35	65
Petuniasterone D (XIII)	484	130	269
12-Acetoxypetuniasterone D Acetate (XIV)	584	135	231

a. ED_{50} is defined as dietary concentration of additive sufficient to reduce growth to 50% of control subjects grown on artificial diet with no additive. Growth period was ten days.

Of these compounds, only those having the bicyclic orthoester system were significantly active in reducing larval growth (Table III). We had originally suspected that the A-ring dienone system

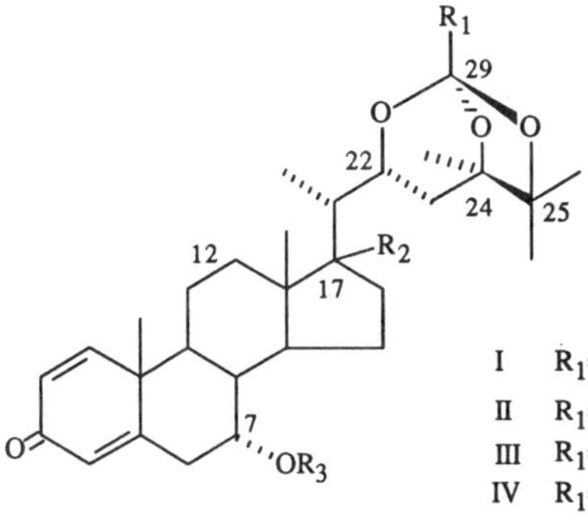

Petuniasterone A Series

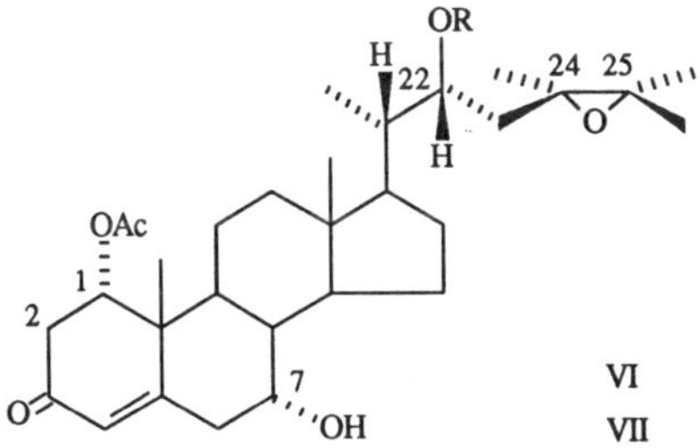

I R_1=CH_2COSCH_3, R_2, R_3=H
II R_1=CH_2COSCH_3, R_2=OH, R_3=H
III R_1=CH_2COSCH_3, R_2=OH, R_3=Ac
IV R_1=CHOHCOSCH_3, R_2, R_3=H
V R_1=CHOHCOSCH_3, R_2=H, R_3=Ac

Petuniasterone B Series

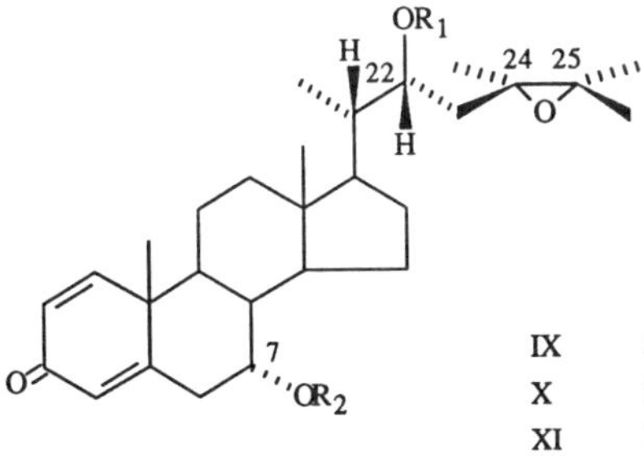

VI R=$COCH_2COSCH_3$
VII R=Ac
VIII R=H

Petuniasterone C Series

IX R_1=$COCH_2COSCH$, R_2=H
X R_1=Ac, R_2=H
XI R_1=H, R_2=H
XII R_1=H, R_2=Ac

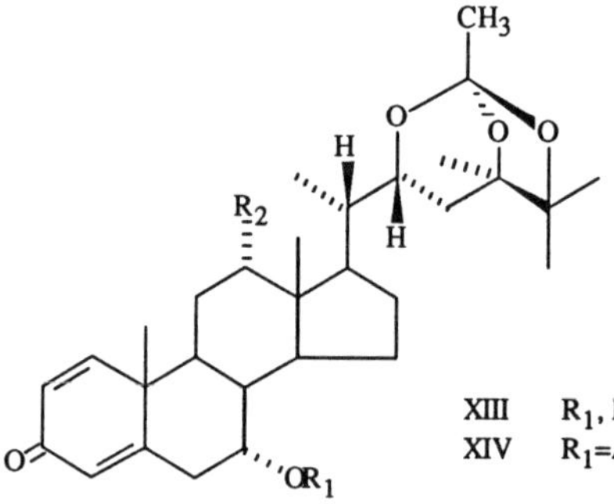

XIII R_1, R_2=H
XIV R_1=Ac, R_2=OAc

Petuniasterone D Series

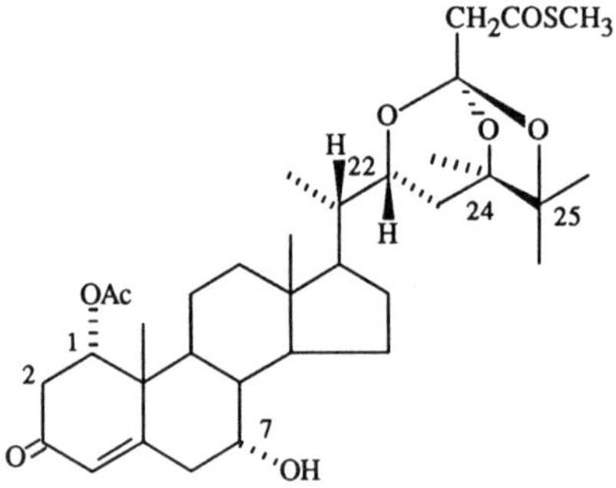

XV Petuniasterone E

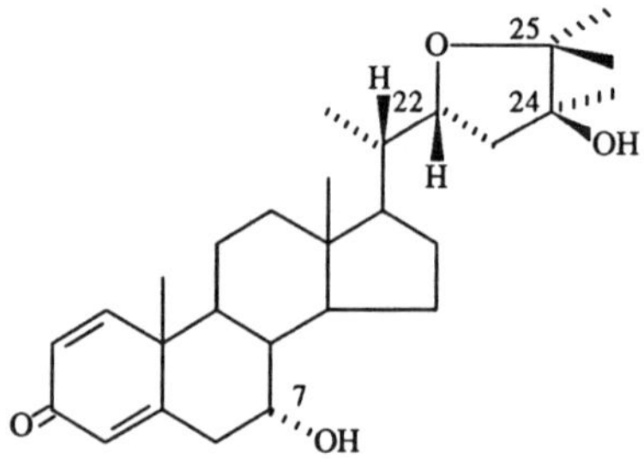

XVI Petuniasterone F

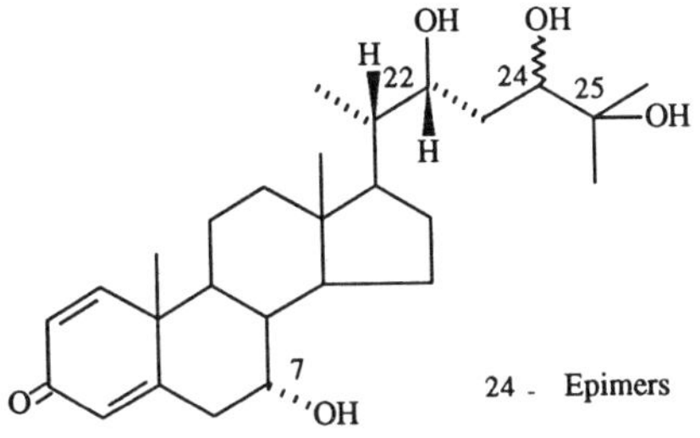

24 - Epimers

XVII Petuniasterones G_1 & G_2

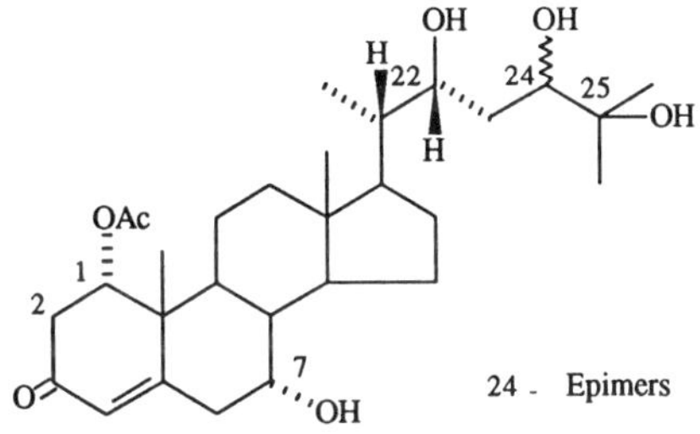

24 - Epimers

XVIII Petuniasterones H_1 & H_2

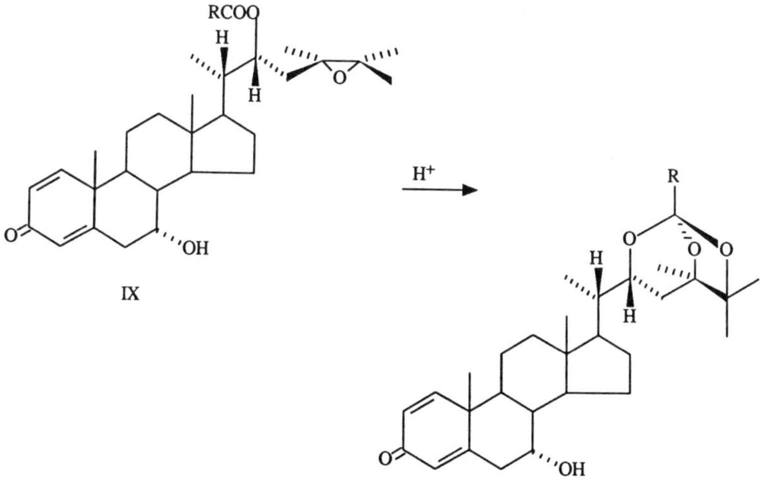

IX

H^+

I

(1)

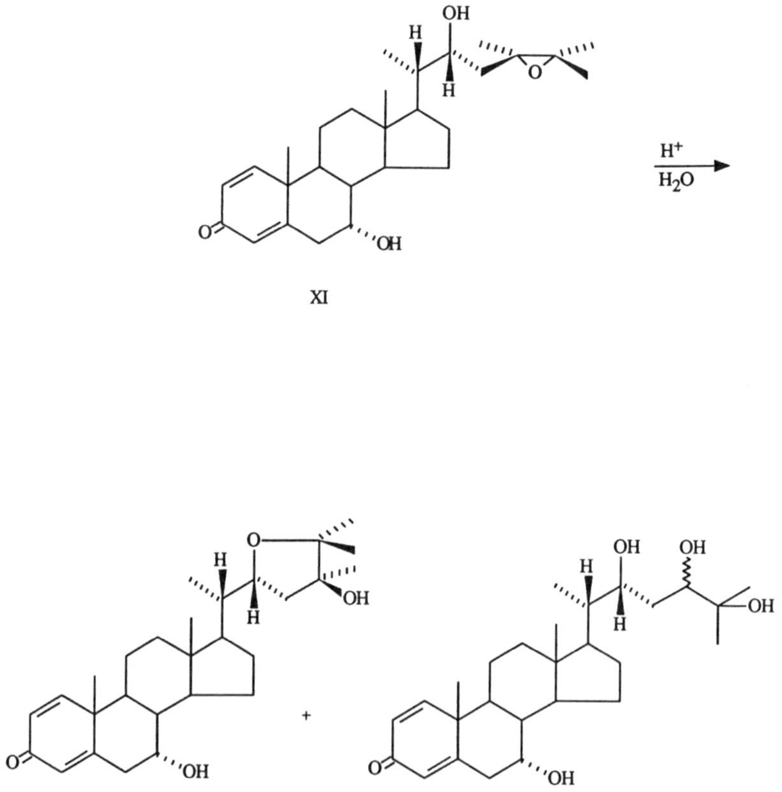

XI

$\xrightarrow[H_2O]{H^+}$

XVI + XVII (2)

Table IV. Inactive Substances[a]

Compound	Max. Level Tested	Growth at Max. Level[b]
All Petuniasterone B (VI-VIII)	800 PPM	70-95%
All Petuniasterone C (IX-XII)	800 PPM	80-94%
Petuniasterone F (XVI)	800 PPM	89%
30-Hydroxy PS-A (IV)	400 PPM	91%
17-Hydroxy PS-A (II)	400 PPM	69%
17-Hydroxy PS-A, Acetate (III)	800 PPM	67%
Testosterone	800 PPM	102%
4-Cholesten-3-one	800 PPM	84%
1-Dehydrotestosterone	800 PPM	104%

a. Materials not reducing growth of *H. zea* larvae to 50% of control values at maximum level tested after ten days.
b. Growth expressed as percent of control wts. for larvae grown on diets without additive.

gave rise to inhibitory activity, but our observations that the C-series compounds (IX-XII) and PS-F (XVI) were inactive showed that this was not true. The model compound, 1-dehydrotestosterone, also having an A-ring dienone actually proved to be mildly stimulatory to growth. Presence of the thiolester moiety was not essential for activity. PS-D (XIII) with its unsubstituted orthoacetate was just as active as PS-A (I) and the methyl ester formed from the thiolester by trans-esterification of PS-A was the most active compound tested. Acetylation of the hydroxyl group at position -7 does not appear to influence activity; however, our preliminary observations on PS-E (XV) indicate that activity is reduced when the A-ring dienone system is not present. Only a small amount of XV was available for testing, and bioassay was conducted on concentrations up to 150 ppm in this case. Larval growth was 69% of control weight at this level. Further substitution of hydroxyl at position -17 (II & III) or at position -30 (IV) of the orthoester appears to eliminate inhibitory activity. All of these observations taken together do not point to a specific mode of action. It is tempting to speculate, however, that some disruptive action upon the insect's hormonal system is occurring because of the steroidal nature of these compounds. In another case involving the effect of diterpene resin acids upon lepidopterans it was shown that cholesterol at high dietary levels could reverse inhibition caused by the acids (16). In the present study, cholesterol at levels of up to 4000 ppm had no effect upon the action of petuniasterone A.

We have noted that for dietary levels of active substances in the range of approximately the ED_{50} most larvae survive even though stunted, whereas for larvae on petunia leaves or on diets containing crude extracts many larvae die. In fresh leaves of Royal Cascade petunia the highest level of PS-A (I) was about 120 ppm and that of PS-D (XIII) and 12-acetoxypetuniasterone D, 7-acetate (XIV) was 30 and 100 ppm respectively. We found that inhibitory activity of PS-A (I) and PS-D (XIII) was additive and not synergistic (this may be true for all active compounds). The activity of PS-A was not changed by the presence of the inactive PS-B, 22-hemithiomalonate (VI), and it may be assumed that the other inactive substances do not significantly enhance or decrease the effect of the inhibitory petuniasterones. The level of active compounds in the petunia cultivar studied is thereby about 250 ppm or approximately twice the ED_{50} and this amount typically reduces larval growth to about 25% of control weights with nearly 100% survival. The greater activity of crude material suggests that additional, significantly active substances have not been isolated.

Among commercial varieties of *Petunia hybrida*, the content of petuniasterones is quite variable. We have observed, for example, that the variety "Blue Jeans" has almost no PS-A in its leaves. It was of interest to examine the putative ancestors of the hybrids both with respect to insect resistance in general and to the various petuniasterones in particular. There is some disagreement concerning the actual species of *Petunia* that have given rise to the present hybrids, but it is felt that *P. violacea*, *P. parodii*, *P. axillaris* and *P. inflata* are likely ancestors (*17*). We were able to acquire seed of all but the last mentioned and grow plants for evaluation. Table V compares larval survival on leaves of these three *Petunia* species with that of larvae on two commercial varieties and on leaves of a commercial tomato. The difference between plants is impressive and suggests that there may be a qualitative difference for the toxic agents contained within *P. parodii*. This difference is confirmed by comparison of the h.p.l.c. profiles for extracts of these petunias (Figure 1). One can easily see that the insect susceptible *P. violacea* has a very low content of petuniasterones whereas *P. axillaris* more nearly resembles the hybrid already studied. The highly resistant *P. parodii* appears to contain different compounds. We have performed preliminary workups of *P. parodii* and *axillaris* and have isolated several new petuniasterones from the former species, only one of which was abundant within extracts of the latter. Our proposed structures are shown here (XIX-XXIII), but in certain cases the assignment of stereochemistry is not rigorously established. It is

Table V. Survival of 3-Day Old Larvae of *H. zea* upon *Petunia* and *Lycopersicon* Leaves

Plant	Number of Surviving Larvae[a] one day	two days
L. esculentum (Ace)	15	15
P. hybrida (Blue Cloud)	7	3
P. hybrida (Royal Cascade)	6	1
P. violacea	15	13
P. parodii	2	0
P. axillaris	7	3

a. 15 Larvae applied initially.

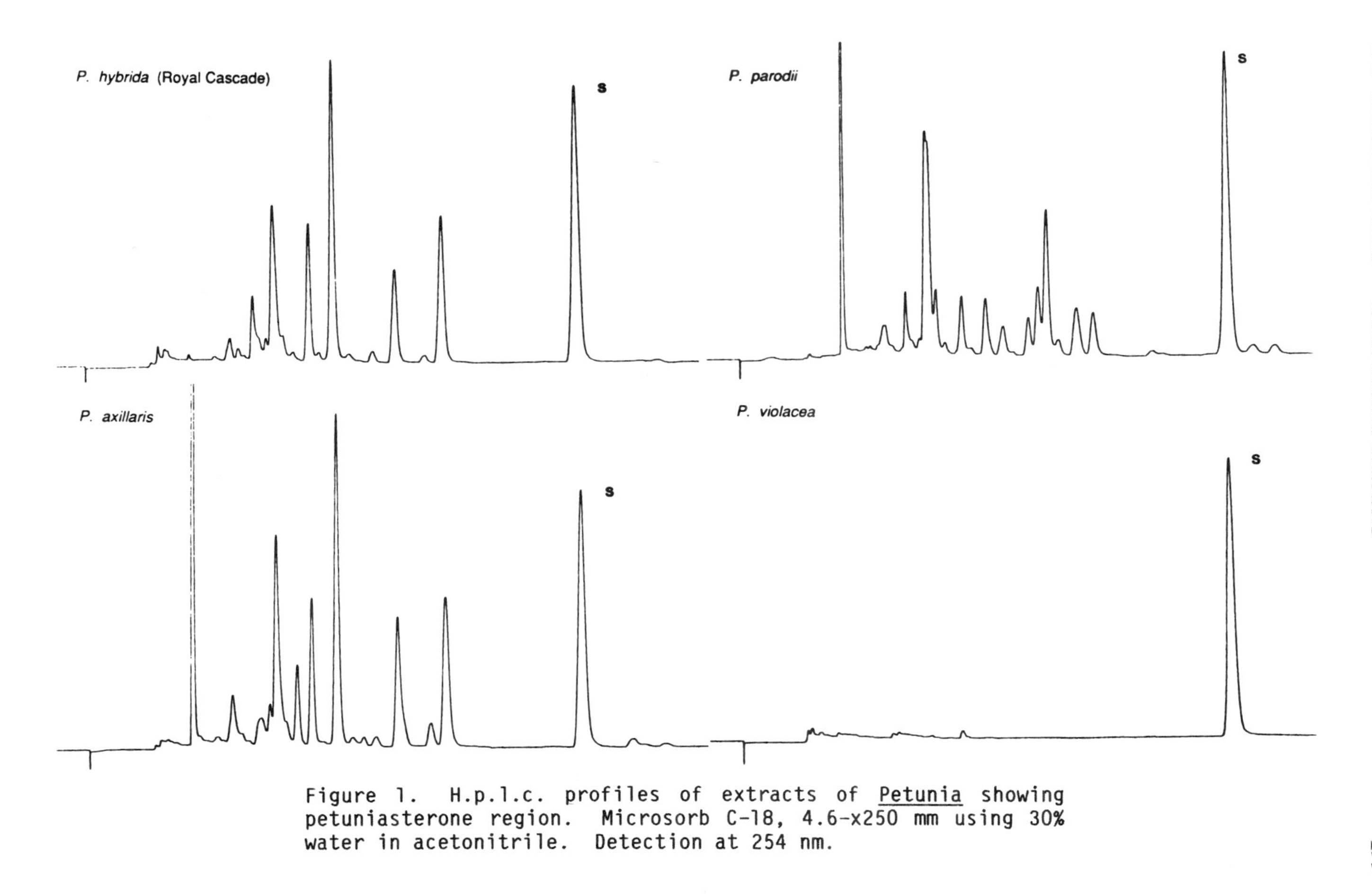

Figure 1. H.p.l.c. profiles of extracts of Petunia showing petuniasterone region. Microsorb C-18, 4.6-x250 mm using 30% water in acetonitrile. Detection at 254 nm.

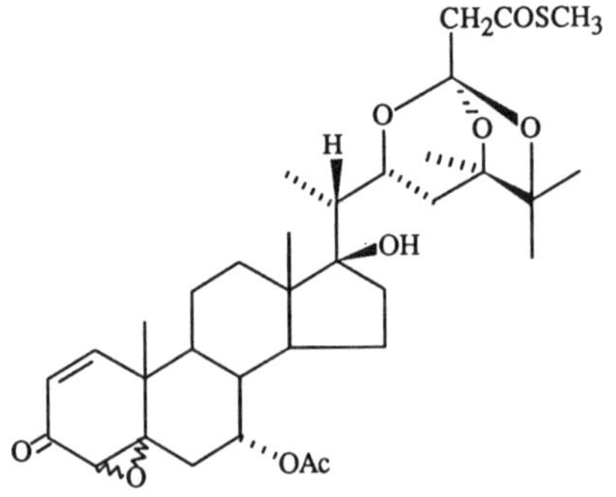

CH_2COSCH_3

OH

OAc

XIX

CH_3

R_2

OR_1

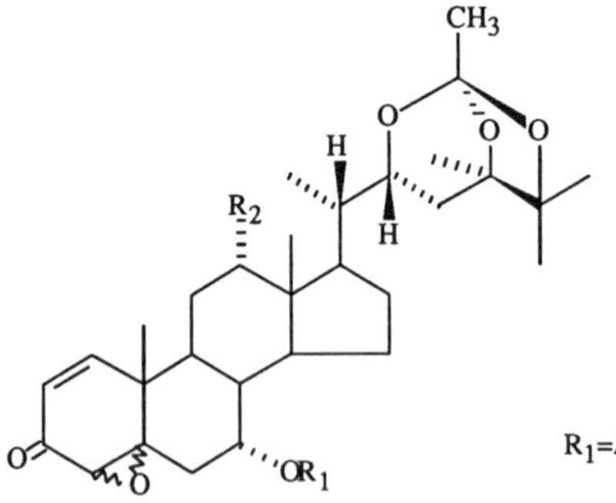

R_1=Ac, R_2=OAc

XX

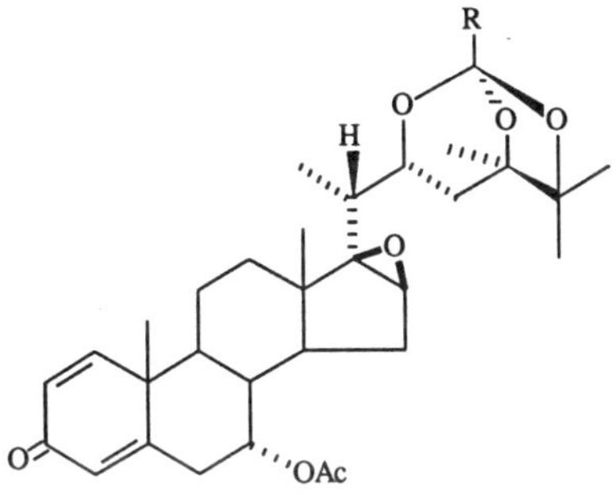

R
O O O
H
O
O
OAc

XXI R=CH_3

XXII R=CH_2COSCH_3

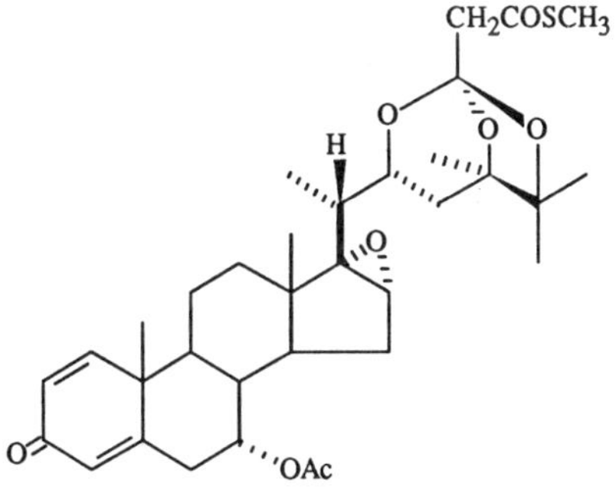

CH_2COSCH_3
O O O
H
O
O
OAc

XXIII

interesting that although we have not yet performed insect bioassays on pure substances the increased insecticidal activity of this plant is correlated with further oxygenation of the petuniasterone system. Also of note, and perhaps not surprising in view of the enhanced toxicity of the plant is that the rapidity of action of crude extracts permits observation of acute effects upon the larvae treated. We have seen rapid onset of uncontrolled diuresis for these larvae after consumption of only a few bites of artificial diet containing these extracts. The animals immediately stop feeding, lose body turgor and become moribund due to dehydration. Such action makes the compounds of P. parodii of special interest to the insect physiologist. From a standpoint of plant breeding it is ironic that artificial selection for color (from violacea) may actually have resulted in a loss of insect resistance for the resulting hybrids which formerly had been present in the white flowered Petunia parodii.

Table VI. Growth of H. zea[a] on Artificial Diets Containing P. peruviana Extracts [b]

Hexane	102 ±6
Ethyl acetate	30 ±6
Acetone	9 ±5
Methanol	0.6 ±0.6

a. Expressed as % of control wts. ± s.d. after ten days.
b. Extracts were incorporated into artificial diets on a wt. basis equivalent to one-fifth the amount originally present in fresh plant leaves.

Inhibitory agents of Physalis. Growth inhibition of H. zea larvae by Physalis peruviana extracts is shown in Table VI. The methanol and acetone fractions were examined further by initial partitioning of their constituents between water and the macroreticular, nonionic resin XAD-2. The insect inhibitory activity of both fractions was associated with material that underwent sorption to the resin. Subsequent chromatographic separations of the materials released from the XAD-2 by desorption with methanol were carried out on Sephadex LH-20 (Methanol as eluent) to give in each case broad zones showing biological activity. Further chromatographic examination by h.p.l.c. using Dynamax C-18 columns with acetonitrile/water gradients showed that considerable overlap of components occurred between the acetone and methanol extracts. The n.m.r. spectra of the still crude mixtures indicated the presence of a number of substances within the withanolide family, a group of steroidal lactones long known to be present in Physalis species as well as in other genera within the Solanaceae (18, 19). Further chromatographic workup on preparative Dynamax C-18 of the LH-20 active zones gave a number of pure or highly enriched substances having a range of biological potency from completely inactive to highly inhibitory at dietary levels below 100 ppm. Most of these compounds remain unidentified, but the ^{1}H n.m.r. and ^{13}C n.m.r. spectra show that aglycones, mono-, di-, and triglycosides are

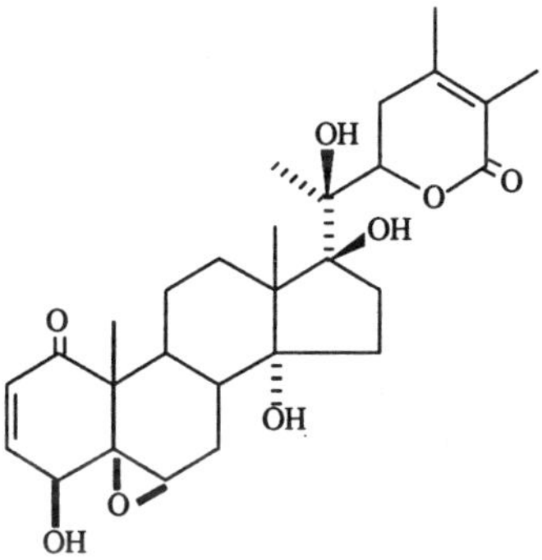

OH
OH
O
O
O
OH
O
OH

XXIV

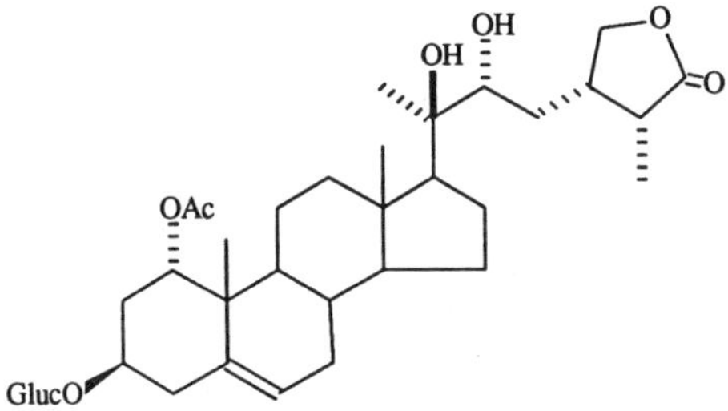

OH
OH
O
O
OAc
GlucO

XXV

represented with the most polar of these occurring in the methanol extract. Two substances, XXIV and XXV have been identified. Compound XXIV, 4-ß-hydroxy withanolide E, has already been reported to have anti-insect activity (20, 21) against larvae of the lepidopteran, Spodoptera littoralis. Our preliminary results indicate an ED_{50} of ca. 250 ppm for XXIV toward H. zea. The 3-ß-glucoside of perulactone (XXV) has not been previously reported although the, aglucone has been described (22). Initial results shown an ED_{50} of 150 ppm (ca. 220 micromolal) in H. zea bioassays. Several of the more highly glycosylated compounds appear to be at least an order of magnitude more active. It is possible that dietary uptake and transport within the insect are enhanced by the increased water solubility of these derivatives.

Conclusion

The purpose of this investigation has been to facilitate the transfer of genetic characteristics related to insect resistance into plants having economic usefulness. The source of this genetic material can profitably lie within plants which are distantly related now that modern methods exist to carry out intergeneric hybridization. The role of the chemist in this process still consists in isolation and structure determination of biologically active substances. In this instance chemistry serves to provide genetic markers for analyses of hybrid progeny. Here the chemist is concerned with the eventual delivery system as well as with the individual components that contribute to insect resistance. At the present time we are conducting experiments on protoplast fusion which should lead to production of many hundreds of plant types. It can be seen that chemical analysis is much more practical for this large number than is recourse to artificial infestation by insect pests, etc.

We expect the eventual intergenerically transformed tomato plants (e.g. Lycopersicon x Physalis) to have a longer lasting or more durable resistance in the face of evolutionary change of the insect pests. Insects that have become adapted to or tolerant toward resistance factors of tomato may not develop detoxification mechanisms for the Physalis chemicals or those of Petunia. Multigenic resistance in plants would have to be met by more complex biochemical evolution for a given pest insect. Evolution of tolerance toward pesticides by insects is rapid as we all know; even such an advanced agent such as Bacillus thurengensis toxin can lose effectiveness under certain circumstances (23,24). We hope to avoid this problem by presenting the pest with an array of phytochemicals that will place a more severe burden upon the evolutionary capability of insects. Thus, multigenic resistance factors will require multiple enzyme systems for detoxification and will place an unacceptable load upon the insect.

Literature Cited

1. Jacobson, M., Botanical Pesticides: Past, Present and Future; This Volume.
2. Maxwell, F. G.; Lafever, H. N.; Jenkins, J. N. J. Econ. Entomol., 1966, 59, 585-588.

3. Metcalf, R. L.; J. Chem. Ecol. 1986, 12, 1109-1124.
4. Juvik, J. A.; Stevens, M. A. J. Amer. Soc. Hort. Sci. 1982, 107, 1061-1065.
5. Maxwell, F. G.; Jennings, P. R., Eds. Breeding Plants Resistant to Insects; Wiley-Interscience: New York, 1980; 683p.
6. Gibson, R. W. Amer. Potato J. 1978, 55, 595-599.
7. Elliger, C. A.; Wong, Y.; Chan, B. G.; Waiss, A. C., Jr. J. Chem. Ecol. 1981, 7, 753-758.
8. Isman, M. B.; Duffey, S. S. Entomol. Exp. Appl. 1982, 31, 370-376.
9. Campbell, B. C.; Duffey, S. S. J. Chem. Ecol. 1981, 7, 927-946.
10. Tanksley, S. D.; Bernatzky, R. Molecular Markers for the Nuclear Genome of Tomato, In Tomato Biotechnology; ARL: New York, 1987; pp 37-44.
11. Shaver, T. N.; Parrott, W. L. J. Econ. Entomol. 1970, 63, 1802-1804.
12. Rick, C. M. Genetic Resources in Lycopersicon, In Tomato Biotechnology; ARL: New York, 1987; pp 17-26.
13. Williams, W. G.; Kennedy, G. G.; Yamamoto, R. T.; Thacker, J. D.; Bordner, J. Science 1980, 207, 777-889.
14. Chan, B. G.; Waiss, A. C., Jr.; Stanley, W. L.; Goodban, A. E. J. Econ. Entomol. 1978, 71, 366-368.
15. Elliger, C. A.; Benson, M. E.; Haddon, W. F.; Lundin, R. E.; Waiss, A. C., Jr.; Wong, R. J. Chem. Soc. Perkin Trans. 1 1988, 711-717.
16. Elliger, C. A.; Zinkel, D. F.; Chan, B. G.; Waiss, A. C. Jr.; Experentia 1976 32, 1364-1365.
17. Sink, K. C. in Petunia; Sink, K. C., Ed.; Springer: Berlin 1984; pp. 3-9.
18. Kirson, I.; Glotter, E. J. Nat. Prod. 1981, 44, 633-647.
19. Glotter, E.; Kirson, I.; Lavie, D.; Abraham, A. In Bioorganic Chemistry; Van Tamelen, E. E., Ed.; Academic Press: New York, 1978; Vol. II pp. 57-95.
20. Ascher, K. R. S.; Nemny, N. E.; Eliyahu, M.; Kirson, I.; Abraham, A.; Glotter, E. Experentia 1980, 36, 998-999.
21. Ascher, K. R. S.; Eliyahu, M.; Glotter, E.; Goldman, A.; Kirson, I.; Abraham, A.; Jacobson, M.; Schmutterer, H. Phytoparasitica 1987, 15, 15-29.
22. Gottlieb, H. G.; Kirson, I.; Glotter, E.; Ray, A.; Sahai, M.; Ali, A. J. Chem. Soc. Perkin Trans. I 1980, 2700-2704.
23. McGaughey, W. H. Science 1985, 229, 193-195.
24. McGaughey, W. H.; Beeman, R. W. J. Econ. Entomol. 1988, 81 28-33.

RECEIVED November 2, 1988

Author Index

Affiliation Index

Subject Index

Production by Paula M. Befard
Indexing by Debby Steiner

Elements typeset by Hot Type Ltd., Washington, DC
Printed and bound by Maple Press, York, PA